The Geometry of the
Octonions

The Geometry of the
Octonions

Tevian Dray
Department of Mathematics
Oregon State University, USA

Corinne A Manogue
Department of Physics
Oregon State University, USA

World Scientific

NEW JERSEY · LONDON · SINGAPORE · BEIJING · SHANGHAI · HONG KONG · TAIPEI · CHENNAI

Published by

World Scientific Publishing Co. Pte. Ltd.

5 Toh Tuck Link, Singapore 596224

USA office: 27 Warren Street, Suite 401-402, Hackensack, NJ 07601

UK office: 57 Shelton Street, Covent Garden, London WC2H 9HE

Library of Congress Cataloging-in-Publication Data
Dray, Tevian.
    The geometry of the octonions / Tevian Dray (Oregon State University, USA) & Corinne A. Manogue (Oregon State University, USA).
        pages cm
    Includes bibliographical references and index.
    ISBN 978-9814401814
    1. Cayley numbers (Algebra) 2. Cayley algebras. 3. Nonassociative algebras. 4. Geometry, Algebraic. I. Manogue, Corinne A. II. Title.
    QA252.5.D73 2015
    512'.5--dc23
                            2015003048

British Library Cataloguing-in-Publication Data
A catalogue record for this book is available from the British Library.

In-house Editors: Lai Fun Kwong/V. Vishnu Mohan

Typeset by Stallion Press
Email: enquiries@stallionpress.com

Printed in Singapore

To David and Tony, for getting us started

Preface

This is a book about the octonions, a bigger and better version of the complex numbers, albeit with some subtle properties. Bigger, because there are more square roots of -1. Better, because an octonionic formalism provides natural explanations for several intriguing results in both mathematics and physics. Subtle, because the rules are more complicated; order matters.

Some readers may be familiar with the quaternions, which lie halfway between the complex numbers and the octonions. Originally developed more than 100 years ago to be the language of electromagnetism, an effort that lost out in the end to the use of vector analysis, the quaternions have been reborn as a useful tool for applications as diverse as aeronautical engineering, computer graphics, and robotics. What will the octonions be good for? This authors believe that the octonions will ultimately be seen as the key to a unified field theory in physics. But that is a topic for another day, although hints of this vision can be found here.

This book is intended as an introduction to the octonions. It is not a mathematics text; theorems and proofs (and references!) are few and far between. Nonetheless, the presentation is reasonably complete, with most results supported by at least the outline of the underlying computations.

The only true prerequisite for reading this book is the ability to multiply matrices, and a willingness to follow computational arguments. Familiarity with linear algebra is a plus, up to the level of finding eigenvalues and eigenvectors. And of course comfort with the complex numbers is a must, or rather a willingness to become comfortable with them.

The book is divided into three parts. Part I discusses several different number systems, emphasizing the octonions. Part II is the heart of the book, taking a detailed look at a particular collection of symmetry groups, including orthogonal, unitary, symplectic, and Lorentz groups, all expressed in terms of division algebras, up to and including the octonions. As we demonstrate, octonions provide the language to describe the so-called *exceptional Lie groups*. Finally, Part III contains a rather eclectic collection of applications of the octonions, in both mathematics and physics.

A companion website for the book is available at

http://octonions.geometryof.org

which is (partially) mirrored at

http://physics.oregonstate.edu/coursewikis/GO/bookinfo.

Tevian Dray
Corinne Manogue
Corvallis, OR
October 2014

Acknowledgments

This book has its origins in the research conducted by one of us (Corinne) nearly 30 years ago. She in turn introduced her husband (Tevian) to the octonions more than 20 years ago; we have collaborated on further research in this area ever since. This book reflects Tevian's efforts to understand what Corinne has taught him; although most of the actual writing was done by Tevian, the final product has been shaped every bit as much by Corinne, however indirectly. Rob Wilson has been a part of this collaboration for the last five years, and has shown us that the mathematical structure of the octonions is even richer than we had dreamed, going far beyond the topics presented here. Special mention and thanks are due our students, Jörg Schray, Jason Janesky, Aaron Wangberg, Joshua Kincaid, and Henry Gillow-Wiles, whose published work has been incorporated where appropriate, and to our most recent collaborator, John Huerta.

This book is however dedicated to David Fairlie and Tony Sudbery, Corinne's collaborators and mentors for her early work relating the octonions to string theory. Our ongoing work with the octonions, including this book, is a testament to their vision; they got us started.

This book would not have been possible without the financial support and encouragement of the John Templeton Foundation, for which we are very grateful.

An early draft of the material in Chapter 14 formed the basis for formal and informal seminars at several institutions, including Mount Holyoke and Grinnell Colleges in 2002, and Oregon State University in 2004. Sections 11.5, 12.4, and much of Chapters 13 and 15 are based on the authors' published research, as cited there.

Finally, all figures in this book appeared initially on the authors' website at http://physics.oregonstate.edu/coursewikis/GO with a Creative Commons by-nc-nd license, and are used by permission.

Contents

List of Figures

List of Tables

Chapter 1

Introduction

What number systems allow the arithmetic operations of addition, subtraction, multiplication, and division?

In order to count, we need integers. How high can we count? There are finite number systems—and the universe itself may be finite—but let's assume that we can count to infinity, or, more precisely, that there's no largest number. In order to subtract, we also need zero, as well as negative numbers. As we said, we need the integers, namely

$$\mathbb{Z} = \{\ldots, -2, -1, 0, 1, 2, \ldots\}. \tag{1.1}$$

Integer arithmetic works fine for addition, subtraction, and multiplication, but what about division? Now we need fractions, or, more formally, the rational numbers

$$\mathbb{Q} = \left\{ \frac{a}{b} : a, b \in \mathbb{Z}; b \neq 0 \right\}. \tag{1.2}$$

Now we can indeed do arithmetic. However, we cannot measure the hypotenuse of a right triangle, or the circumference of a circle. For those operations, we need the real numbers, denoted $\mathbb{R}$, which include rational numbers such as 3 or $\frac{1}{2}$, but also irrational numbers such as $\sqrt{2}$ or π.

Are we done yet? No, because there are algebraic equations we still can't solve. The simplest example is

$$x^2 + 1 = 0 \tag{1.3}$$

whose solutions, if any, would be square roots of -1. As you may know, the complex numbers $\mathbb{C}$ consist of both real numbers and real multiples of $i = \sqrt{-1}$. Furthermore, all nth order polynomial equations have exactly n solutions over $\mathbb{C}$ (counting multiplicity). The complex numbers may seem like a mathematical toy; who needs the square root of negative numbers? It turns out that quantum mechanics is an inherently complex theory; complex

1

numbers are an essential ingredient in our current understanding of the world around us.

Are we done yet? Maybe. What are the rules? Even over the complex numbers, the rules are slightly different, as evidenced by the apparent paradox

$$i = \sqrt{-1} = \sqrt{\frac{1}{-1}} = \frac{\sqrt{1}}{\sqrt{-1}} = \frac{1}{i} = -i \qquad (1.4)$$

which forces us to rethink the rules for manipulating square roots. Can we change the rules in other ways?

Yes, we can. First of all, we can drop the requirement that numbers commute with each other. However counterintuitive this may feel, there is good reason to suspect that such numbers could be useful, since there are many physical operations where the order matters. A mathematical example where order matters is matrix multiplication, where in general $AB \neq BA$. As we will see, this choice leads to the quaternions, $\mathbb{H}$.

Can we go further? Yes, by dropping associativity. Why would we do that? Well, matrix multiplication is associative, but the cross product is not. Can you work out $\vec{v} \times \vec{v} \times \vec{w}$? Not without knowing which product to do first! As we will see, this choice leads to the octonions, $\mathbb{O}$.

Can we go further? No; going further requires us to give up division. More precisely, the four algebras $\mathbb{R}$, $\mathbb{C}$, $\mathbb{H}$, and $\mathbb{O}$ are the only ones without zero divisors, that is, nonzero elements whose product is nonetheless zero.

The octonions are thus nature's largest division algebra. This author believes that this largest mathematical structure will ultimately be seen as the key to understanding the basic building blocks of nature, namely the fundamental particles such as electrons and quarks.

The goal of this book is to introduce the reader to the mathematics of the octonions, while offering some hints as to how they might be useful in physics. Let's begin.

PART I
Number Systems

Chapter 2

The Geometry of the Complex Numbers

2.1 Complex Numbers

Begin with the real numbers, $\mathbb{R}$. Add "the" square root of -1; call it i. You have just constructed the complex numbers, $\mathbb{C}$, in the form

$$\mathbb{C} = \mathbb{R} \oplus \mathbb{R}\,i. \tag{2.1}$$

That is, a complex number z is a pair of real numbers (a,b), which is usually written as

$$z = a + bi \tag{2.2}$$

and which can be thought of as either a point in the (complex) plane with coordinates (a, b) or as a vector with components a and b.

2.2 History

Complex numbers were first used in the 16th century in order to solve cubic equations, as there are some cases with real solutions that nonetheless require the use of complex numbers in order to obtain those solutions. The recognition of the complex numbers as an object worthy of study in their own right is usually attributed to Rafael Bombelli, who in 1572 was the first to formalize the rules of complex arithmetic (and also, at the same time, the first to write down the rules for manipulating negative numbers). The term *imaginary* was introduced only later, by René Descartes in 1637.

2.3   Algebra

The complex numbers are more than just a *vector space*; they are also an *algebra*, that is you can multiply them together. How do you compute the product of complex numbers? Simply multiply it out, that is

$$(a + bi)(c + di) = (a + bi)c + (a + bi)di$$
$$= (ac - bd) + (bc + ad)i. \tag{2.3}$$

What properties of the complex numbers have we used? First of all, we have distributed multiplication over addition. Second, i is "the" square root of -1, that is

$$i^2 = -1. \tag{2.4}$$

Third, we have used associativity, that is

$$(xy)z = x(yz) \tag{2.5}$$

for any complex numbers x, y, z. Finally, we have used commutativity, i.e.

$$xy = yx \tag{2.6}$$

to replace bic with bci.[1]

We define the *complex conjugate* $\bar{z}$ of a complex number $z = a + bi$ by

$$\bar{z} = a - bi \tag{2.7}$$

thus changing the sign of the imaginary part of z. Equivalently, complex conjugation is the (real) linear map which takes 1 to 1 and i to $-i$. The *norm* $|z|$ of a complex number z is defined by

$$|z|^2 = z\bar{z} = a^2 + b^2. \tag{2.8}$$

The only complex number with norm zero is zero. Furthermore, any nonzero complex number has a unique inverse, namely

$$z^{-1} = \frac{\bar{z}}{|z|^2}. \tag{2.9}$$

Since complex numbers are invertible, linear equations such as

$$c = az + b \tag{2.10}$$

can always be solved for z, so long as $a \neq 0$.

The norms of complex numbers satisfy the following identity:

$$|yz| = |y||z|. \tag{2.11}$$

Squaring both sides and expanding the result in terms of components yields

$$(ac - bd)^2 + (bc + ad)^2 = (a^2 + b^2)(c^2 + d^2) \tag{2.12}$$

(where, say, $z = a + bi$ and $y = c + di$), which is called the *2-squares rule*.

---

[1] We are really assuming that multiplication is *linear* over the reals. This not only implies distributivity, but also commutativity between real numbers and the complex unit, i, which in this case is enough to ensure full commutativity.

2.4 Geometry

Thanks to *Euler's formula*,

$$e^{i\theta} = \cos\theta + i\sin\theta, \qquad (2.13)$$

polar coordinates can be used to write complex numbers in terms of their norm and a phase angle θ. (A factor of the form $e^{i\theta}$ is called a *phase*.) That is, any complex number can be written in the form

$$z = re^{i\theta} \qquad (2.14)$$

where

$$r = |z| \qquad (2.15)$$

since $|e^{i\theta}| = 1$. Each complex number thus has a direction associated with it in the complex plane, determined by the angle θ.

Euler's formula provides an elegant derivation of the angle addition formulas for sine and cosine. We have

$$(\cos\alpha + i\sin\alpha)(\cos\beta + i\sin\beta) = e^{i\alpha}e^{i\beta} = e^{i(\alpha+\beta)}$$

$$= \cos(\alpha+\beta) + i\sin(\alpha+\beta) \qquad (2.16)$$

so that working out the left-hand side using complex multiplication, and comparing real and imaginary parts, yields the standard formulas for the trigonometric functions on the right-hand side.

We can use (2.16) to provide a geometric interpretation of complex multiplication. We have

$$(r_1 e^{i\theta_1})(r_2 e^{i\theta_2}) = r_1 r_2 e^{i(\theta_1+\theta_2)} \qquad (2.17)$$

so the result of multiplying one complex number, z_1 by another, z_2, is to stretch z_1 by the magnitude r_2 of z_2, and to rotate it counterclockwise by the phase angle θ_2 of z_2. The same product can of course be reinterpreted with the roles of z_1 and z_2 reversed. A special case is that multiplication by i rotates a complex number counterclockwise by $\frac{\pi}{2}$, without changing its norm.

Euler's formula is usually proved by comparing the power series expansions of each side. An alternative proof is obtained by noticing that both sides of this equation satisfy the differential equation

$$\frac{d^2 f}{d\theta^2} = -f$$

with the same initial conditions. A special case of Euler's formula is the famous equation

$$e^{i\pi} + 1 = 0$$

which relates five of the most basic symbols in mathematics!

Chapter 3

The Geometry of the Quaternions

3.1 Quaternions

What happens if we include another, independent, square root of -1? Call it j. Then the big question is, what is ij?

Hamilton eventually proposed that $k = ij$ should be yet another square root of -1, and that the multiplication table should be cyclic, that is

$$ij = k = -ji, \tag{3.1}$$

$$jk = i = -kj, \tag{3.2}$$

$$ki = j = -ik. \tag{3.3}$$

We refer to i, j, and k as *imaginary quaternionic units*. Notice that these units anticommute!

This multiplication table is shown schematically in Figure 3.1. Multiplying two of these quaternionic units together in the direction of the arrow yields the third; going against the arrow contributes an additional minus sign.

The *quaternions* are denoted by $\mathbb{H}$; the "H" is for Hamilton.[1] They are spanned by the identity element 1 and three imaginary units, that is, a quaternion q can be represented as four real numbers (q_1, q_2, q_3, q_4), usually written

$$q = q_1 + q_2 i + q_3 j + q_4 k \tag{3.4}$$

which can be thought of as a point or vector in $\mathbb{R}^4$. Since (3.4) can be written in the form

$$q = (q_1 + q_2 i) + (q_3 + q_4 i)j \tag{3.5}$$

---

[1] The symbol $\mathbb{Q}$ is used to denote the *rational* numbers, and is therefore not available for the quaternions.

9

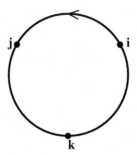

Fig. 3.1   The quaternionic multiplication table.

we see that a quaternion can be viewed as a pair of complex numbers $(q_{1C}, q_{2C}) = (q_1 + q_2 i, q_3 + q_4 i)$, or equivalently that we can write

$$\mathbb{H} = \mathbb{C} \oplus \mathbb{C} j \tag{3.6}$$

in direct analogy to the construction of $\mathbb{C}$ from $\mathbb{R}$.

3.2   History

The quaternions were discovered by Sir William Rowan Hamilton in 1843, after struggling unsuccessfully to construct an algebra in three dimensions. On 16 October 1843, as Hamilton was walking along a canal in Dublin, he realized how to construct an algebra in four dimensions instead. In perhaps the most famous act of mathematical graffiti of all time, Hamilton carved the multiplication table of the quaternions,

$$i^2 = j^2 = k^2 = ijk = -1 \tag{3.7}$$

onto the base of the Brougham Bridge as he passed it. Although the original carving is now gone, a plaque marking this historic event appears in its stead (complete with graffiti!), as shown in Figure 3.2. The inscription on the plaque reads: *Here as he walked by on the 16th of October 1843 Sir William Rowan Hamilton in a flash of genius discovered the fundamental formula for quaternion multiplication $i^2 = j^2 = k^2 = ijk = 1$ & cut it on a stone of this bridge.*

The quaternions provided the first natural language in which to discuss electromagnetism. The modern language of vector analysis was not introduced until the mid-1880s, by Josiah Willard Gibbs, Oliver Heaviside, and Hermann von Helmholtz. After lengthy disagreement, the language of

Fig. 3.2 The Brougham Bridge in Dublin (left), and the plaque there commemorating Hamilton's discovery (right).

vector analysis won out, and to this day electromagnetism is taught almost exclusively using vectorial methods.[2]

However, quaternions have made a comeback in recent years, as they provide a natural language in which to describe spatial rotations. Quaternions are currently used in applications ranging from aeronautics and robotics to video games.

3.3   Algebra

The quaternionic multiplication table is almost, but not quite, the vector cross product. The only difference is that imaginary quaternions square to a negative number, whereas the cross product of a vector with itself is zero.

This is not a coincidence. Making the obvious identification of vectors $\vec{v}$, $\vec{w}$ with imaginary quaternions v, w, namely

$$\vec{v} = v_x\hat{\imath} + v_y\hat{\jmath} + v_z\hat{k} \longleftrightarrow v = v_x i + v_y j + v_x k \qquad (3.8)$$

(and similarly for $\vec{w}$), then the imaginary part of the quaternionic product vw is the cross product $\vec{v} \times \vec{w}$, that is

$$\vec{v} \times \vec{w} \longleftrightarrow \mathrm{Im}\,(vw) \qquad (3.9)$$

while the real part is just (minus) the dot product $\vec{v} \cdot \vec{w}$, that is

$$-\vec{v} \cdot \vec{w} = \mathrm{Re}\,(vw). \qquad (3.10)$$

Thus, the quaternionic product can be thought of as a combination of the dot and cross products! In fact, the use of $\hat{\imath}$, $\hat{\jmath}$, $\hat{k}$ for Cartesian basis vectors originates with the quaternions, which predate the use of vectors [2].

---

[2]A noteworthy exception is the book by Baylis [1].

We define the *commutator* of two quaternions p and q by

$$[p, q] = pq - qp \tag{3.11}$$

which quantifies the lack of commutativity of the quaternions. For example, we have $[i, j] = 2k$. However, the quaternions are associative; it is sufficient to check that

$$(ij)k = -1 = i(jk). \tag{3.12}$$

As always, we have distributivity of multiplication over addition.

The (quaternionic) *conjugate* $\bar{q}$ of a quaternion q is obtained via the (real) linear map which reverses the sign of each imaginary unit, so that

$$\bar{q} = q_1 - q_2 i - q_3 j - q_4 k \tag{3.13}$$

if q is given by (3.4). Conjugation leads directly to the *norm* of a quaternion $|q|$, defined by

$$|q|^2 = q\bar{q} = q_1^2 + q_2^2 + q_3^2 + q_4^2. \tag{3.14}$$

Again, the only quaternion with norm zero is zero, and every nonzero quaternion has a unique inverse, namely

$$q^{-1} = \frac{\bar{q}}{|q|^2}. \tag{3.15}$$

Quaternionic conjugation satisfies the identity

$$\overline{pq} = \bar{q}\,\bar{p} \tag{3.16}$$

from which it follows that the norm satisfies

$$|pq| = |p||q|. \tag{3.17}$$

Squaring both sides and expanding the result in terms of components yields the *4-squares rule*,

$$(p_1 q_1 - p_2 q_2 - p_3 q_3 - p_4 q_4)^2 + (p_2 q_1 + p_1 q_2 - p_4 q_3 + p_3 q_4)^2$$
$$+ (p_3 q_1 + p_4 q_2 + p_1 q_3 - p_2 q_4)^2 + (p_4 q_1 - p_3 q_2 + p_2 q_3 + p_1 q_4)^2$$
$$= (p_1^2 + p_2^2 + p_3^2 + p_4^2)(q_1^2 + q_2^2 + q_3^2 + q_4^2) \tag{3.18}$$

which is not quite as obvious as the 2-squares rule. This identity implies that the quaternions form a *division algebra*, that is, not only are there inverses, but there are no zero divisors—if a product is zero, one of the factors must be zero.

Since quaternions are invertible, linear equations such as (2.10), where now $a, b, c, z \in \mathbb{H}$, can still be solved for z so long as $a \neq 0$. But these are no longer the only linear equations! Consider for instance the equation

$$d = ax + xb \tag{3.19}$$

where now $a, b, d \in \mathbb{C}$ and $x \in \mathbb{H}$. One solution is clearly

$$x_0 = \frac{d}{a+b} \tag{3.20}$$

provided $a + b \neq 0$. Note that x_0 is complex. Are there any other solutions? Consider the special case $d = 0$, $a = i = b$. Then *any* linear combination of j and k solves the equation!

This turns out to be the generic situation: If $a + \bar{b} \neq 0$, then the only solution is the complex solution (3.20), but if $a + \bar{b} = 0$ there are additional quaternionic "homogeneous" solutions, which can be added to the particular solution x_0, which is therefore not unique. The situation rapidly becomes more complicated if some or all of a, b, d are themselves allowed to be quaternionic, rather than complex.

3.4 Geometry

It is important to realize that $\pm i$, $\pm j$, and $\pm k$ are not the only square roots of -1. Rather, *any* imaginary quaternion squares to a negative number, so it is only necessary to choose its norm to be one in order to get a square root of -1. The imaginary quaternions of norm one form a sphere; in the above notation, this is the set of points

$$q_2^2 + q_3^2 + q_4^2 = 1 \tag{3.21}$$

(with $q_1 = 0$). Any such unit imaginary quaternion u can be used to construct a complex subalgebra of $\mathbb{H}$, which we will also denote by $\mathbb{C}$, namely

$$\mathbb{C} = \{a + b\,u\} \tag{3.22}$$

with $a, b \in \mathbb{R}$. Furthermore, we can use the identity (2.13) to write

$$e^{u\theta} = \cos\theta + u\sin\theta. \tag{3.23}$$

This means that *any* quaternion can be written in the form

$$q = re^{u\theta} \tag{3.24}$$

where

$$r = |q| \tag{3.25}$$

and where u denotes the *direction* of the imaginary part of q.

A useful strategy for solving problems such as the linear equations in the previous section, which involve both complex numbers and quaternions, is to break up the quaternions into a pair of complex numbers. Consider the following examples.

We define *conjugation*[3] of one quaternion q by another quaternion p by pqp^{-1}. The norm of p is irrelevant here, so we might as well assume that $|p| = 1$, in which case $p^{-1} = \bar{p}$

- *What is the result of conjugating a quaternion by i?*

Write q in terms of a pair of complex numbers via

$$q = q_{1c} + q_{2c}j. \tag{3.26}$$

Then i commutes with the complex numbers q_{1c} and q_{2c}, but anticommutes with j. Thus,

$$i q \bar{\imath} = i q_{1c} \bar{\imath} + i q_{2c} j \bar{\imath} = i q_{1c} \bar{\imath} - i q_{2c} \bar{\imath} j = q_{1c} - q_{2c} j. \tag{3.27}$$

Conjugation by i therefore leaves the complex plane untouched, but yields a rotation by π in the jk-plane. Analogous results would hold for conjugation by any other imaginary quaternionic unit, such as u.

- *What is the result of conjugating a quaternion by $e^{i\theta}$?*

Interchanging the roles of i and j in the previous discussion, conjugation by j yields a rotation by π in the ki-plane, so that

$$j e^{-i\theta} \bar{\jmath} = e^{i\theta}. \tag{3.28}$$

Multiplying both of these equations on the right by j yields the important relation

$$j e^{-i\theta} = e^{i\theta} j. \tag{3.29}$$

Thus,

$$\begin{aligned}
e^{i\theta} q e^{-i\theta} &= e^{i\theta} q_{1c} e^{-i\theta} + e^{i\theta} q_{2c} j e^{-i\theta} \\
&= e^{i\theta} q_{1c} e^{-i\theta} + e^{i\theta} q_{2c} e^{i\theta} j \\
&= q_{1c} + q_{2c} e^{2i\theta} j
\end{aligned} \tag{3.30}$$

corresponding to a rotation by 2θ in the jk-plane. We will return to such examples when discussing symmetry groups in Chapter 6.

---

[3] *Conjugation by p is quite different from* conjugation of p (which would be $\bar{p}$).

Chapter 4

The Geometry of the Octonions

4.1 Octonions

What happens if we include another, independent, square root of -1? Call it ℓ. Here we go again.

In analogy to the previous construction of $\mathbb{C}$ and $\mathbb{H}$, an octonion x can be thought of as a pair of quaternions, $(x_{1\mathbb{H}}, x_{2\mathbb{H}})$, so that

$$\mathbb{O} = \mathbb{H} \oplus \mathbb{H}\,\ell. \tag{4.1}$$

Since we are running out of letters, we will denote i times ℓ simply as $i\ell$, and similarly with j and k. But what about the remaining products?

Of course, $\ell^2 = -1$; this is built into the construction. It is easy to see that $i\ell$, $j\ell$, and $k\ell$ also square to -1; there are now seven independent imaginary units, and we could write

$$x = x_1 + x_2 i + x_3 j + x_4 k + x_5 k\ell + x_6 j\ell + x_7 i\ell + x_8 \ell \tag{4.2}$$

where $x_m \in \mathbb{R}$, which can be thought of as a point or vector in $\mathbb{R}^8$. The *real part* of x is just x_1; the *imaginary part* of x is everything else. Algebraically, we could define

$$\mathrm{Re}\,(x) = \frac{1}{2}(x + \overline{x}), \tag{4.3}$$

$$\mathrm{Im}\,(x) = \frac{1}{2}(x - \overline{x}), \tag{4.4}$$

where it is important to note that the imaginary part is, well, imaginary. This differs slightly from the standard usage of these terms for complex numbers, where "$\mathrm{Im}\,(z)$" normally refers to a real number, the coefficient of i. This convention is not possible here, since the imaginary part has seven degrees of freedom, and can be thought of as a vector in $\mathbb{R}^7$.

The full multiplication table is summarized in Figure 4.1 by means of the 7-point projective plane. Each point corresponds to an imaginary unit.

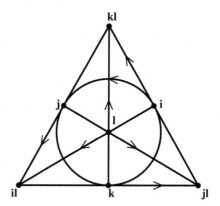

Fig. 4.1   The octonionic multiplication table.

Each line corresponds to a quaternionic triple, much like $\{i, j, k\}$, with the arrow giving the orientation. For example,

$$k\,\ell = k\ell, \tag{4.5}$$
$$\ell\,k\ell = k, \tag{4.6}$$
$$k\ell\,k = \ell, \tag{4.7}$$

and each of these products anticommutes, that is, reversing the order contributes a minus sign.

4.2   History

Remember the Brougham Bridge, where Hamilton carved the quaternionic multiplication table in 1843? Well, on 21 July 2004, as one of the authors walked along that very same canal in Dublin... (See Figure 4.2.)

No, that's not how the octonions were discovered. Hamilton sent a letter to his good friend John T. Graves the day after his discovery in October 1843. On 26 December 1843, Graves wrote back describing the octonions, which he called *octaves*, a name that is still sometimes used. However, Graves didn't publish this work until 1845, shortly after (and in response to) Arthur Cayley's publication of his own discovery of the octonions. For this reason, the octonions are also known as *Cayley numbers*. Although Hamilton later vouched for Graves' priority, Cayley did publish first; both are given credit for independently discovering the octonions.

Fig. 4.2   One of the authors at the Brougham Bridge in Dublin.

4.3   Algebra

We define the *octonionic conjugate* $\overline{x}$ of an octonion x as the (real) linear map which reverses the sign of each imaginary unit. Thus,

$$\overline{x} = x_1 - x_2 i - x_3 j - x_4 k - x_5 k\ell - x_6 j\ell - x_7 i\ell - x_8 \ell \qquad (4.8)$$

if x is given by (4.2). Direct computation shows that

$$\overline{xy} = \overline{y}\,\overline{x}. \qquad (4.9)$$

The *norm* of an octonion $|x|$ is defined by

$$|x|^2 = x\overline{x} = x_1^2 + x_2^2 + x_3^2 + x_4^2 + x_5^2 + x_6^2 + x_7^2 + x_8^2. \qquad (4.10)$$

Again, the only octonion with norm zero is zero, and every nonzero octonion has a unique inverse, namely

$$x^{-1} = \frac{\overline{x}}{|x|^2}. \qquad (4.11)$$

As with the other division algebras, the norm satisfies the identity

$$|xy| = |x||y|. \qquad (4.12)$$

Writing out this expression in terms of components yields the *8-squares rule*, which is no longer at all obvious. The octonions therefore also form a *division algebra*.

A remarkable property of the octonions is that they are *not* associative! For example, compare

$$(i\,j)(\ell) = +(k)(\ell) = +k\ell, \qquad (4.13)$$

$$(i)(j\,\ell) = (i)(j\ell) = -k\ell. \qquad (4.14)$$

However, the octonions are *alternative*, that is, products involving no more than two independent octonions do associate. Specifically,

$$(xy)y = xy^2, \tag{4.15}$$

$$(xy)x = x(yx) \tag{4.16}$$

for any octonions x, y. Alternativity extends to products with conjugates, so that

$$(xy)\bar{y} = x|y|^2, \tag{4.17}$$

$$(xy)\bar{x} = x(y\bar{x}). \tag{4.18}$$

The commutator (3.11), defined for quaternions, extends naturally to the octonions as well. The *associator* of three octonions x, y, z is given by

$$[x, y, z] = (xy)z - x(yz) \tag{4.19}$$

which quantifies the lack of associativity. Alternativity can be phrased as

$$[x, y, x] = 0 = [x, y, y]. \tag{4.20}$$

More generally, both the commutator and associator are *antisymmetric*, that is, interchanging any two arguments changes the result by a minus sign; replacing any argument by its conjugate has the same effect, because the real parts don't contribute.

A consequence of alternativity is that the *Moufang identities*

$$(xyx)z = x\big(y(xz)\big), \tag{4.21}$$

$$z(xyx) = \big((zx)y\big)x, \tag{4.22}$$

$$(xy)(zx) = x(yz)x, \tag{4.23}$$

are satisfied, all of which follow from the associator identity

$$[x, y, zx] = x[x, y, z]. \tag{4.24}$$

Polarizing (4.10), we obtain an inner product on $\mathbb{R}^8$, namely

$$p \cdot q = \frac{1}{2}\big(|p + q|^2 - |p|^2 - |q|^2\big) = p\bar{q} + q\bar{p} = \bar{p}q + \bar{q}p \tag{4.25}$$

where the last two equalities make use of the properties of the commutator, which imply that $[p, \bar{q}] = [\bar{p}, q]$. If v and w are imaginary octonions, then

$$-v \cdot w = vw + wv = \{v, w\} \tag{4.26}$$

where the last equality defines the *anticommutator* of v and w. Thus, two imaginary octonions are *orthogonal* if and only if their anticommutator vanishes. Furthermore, if v is an imaginary octonion, and x is orthogonal to v, then $x\overline{v} + v\overline{x} = 0$, that is

$$x \perp v \Longrightarrow xv = v\overline{x} \tag{4.27}$$

for $x \in \mathbb{O}$ and $v \in \mathrm{Im}\,\mathbb{O}$, since $\overline{v} = -v$.

The analogy with vectors in three dimensions discussed in Section 3.3, also holds in seven dimensions. Identifying v, w with $\vec{v}$, $\vec{w}$, not only is there a dot product, given either by (4.26) or equivalently by (3.9), but there is also a cross product, defined by (3.10). The dot product exists in all dimensions, but the cross product does not. Which way should the cross product point in higher dimensions, where the "right-hand rule" isn't sufficient? The quaternionic and octonionic multiplication table provide an answer to this question, and it turns out that these are the only possibilities: The cross product exists only in three and seven dimensions!

4.4 Geometry

As with the quaternions, the units i, j, k, $k\ell$, $j\ell$, $i\ell$, and ℓ are by no means the only square roots of -1. Rather, *any* imaginary octonion squares to a negative number, so it is only necessary to choose its norm to be one in order to get a square root of -1. The imaginary octonions of norm one form a 6-sphere in the 7-dimensional space of imaginary octonions.

Any such unit imaginary octonion $\widehat{s}$ can be used to construct a complex subalgebra of $\mathbb{O}$, which we will also denote by $\mathbb{C}$, and which takes the form

$$\mathbb{C} = \{a + b\,\widehat{s}\} \tag{4.28}$$

with $a, b \in \mathbb{R}$. We can again use the identity (2.13) to write

$$e^{\widehat{s}\theta} = \cos\theta + \widehat{s}\sin\theta \tag{4.29}$$

so that *any* octonion can be written in the form

$$x = r e^{\widehat{s}\theta} \tag{4.30}$$

where

$$r = |x| \tag{4.31}$$

Two unit imaginary octonions $\widehat{s}$ and $\widehat{t}$ which point in different directions (this excludes $\widehat{t} = -\widehat{s}$) determine a *quaternionic* subalgebra of $\mathbb{O}$, which we also denote by $\mathbb{H}$, and which takes the form

$$\mathbb{H} = \{a + b\,\widehat{s} + c\,\widehat{t} + d\,\widehat{s}\widehat{t}\} \tag{4.32}$$

where $a, b, c, d \in \mathbb{R}$.

An important technique when working with the octonions is to work with what we call *generic* octonions. Any single octonion x can be assumed to lie in $\mathbb{C}$, that is

$$x = x_1 + x_2 i. \tag{4.33}$$

A second octonion y lies in $\mathbb{H}$, but adds just one new direction, that is

$$y = y_1 + y_2 i + y_3 j \tag{4.34}$$

while a third can be assumed to take the form

$$z = z_1 + z_2 i + z_3 j + z_4 k + z_8 \ell \tag{4.35}$$

that is, a general element of $\mathbb{H}$ plus one further direction. Only with four or more octonions is it necessary to use the full eight dimensions of $\mathbb{O}$! This approach makes it obvious that any products involving no more than two distinct octonions (or their conjugates) must associate—they lie in $\mathbb{H}$! But this is just alternativity!

It is often useful to consider octonions as vectors in $\mathbb{R}^8$. The norm of an octonion is precisely the same as its vector norm. But the dot product can be obtained from the norm by a process known as polarization, as follows. If $\vec{v}$, $\vec{w}$ are vectors in $\mathbb{R}^n$, we have

$$\begin{aligned}
|\vec{v} + \vec{w}|^2 &= (\vec{v} + \vec{w}) \cdot (\vec{v} + \vec{w}) \\
&= \vec{v} \cdot \vec{v} + 2\vec{v} \cdot \vec{w} + \vec{w} \cdot \vec{w} \\
&= |\vec{v}|^2 + |\vec{w}|^2 + 2\vec{v} \cdot \vec{w} \tag{4.36}
\end{aligned}$$

which can be solved for the last term. By analogy, the *dot product* of two octonions is given by (4.25). For imaginary octonions, orthogonality is equivalent to anticommutativity, that is,

$$x \cdot y = 0 \iff x\,y = -y\,x \qquad (x, y \in \operatorname{Im} \mathbb{O}) \tag{4.37}$$

which in turn implies that

$$\overline{xy} = \overline{y}\,\overline{x} = y\,x = -x\,y \tag{4.38}$$

so that orthogonality of imaginary octonions is also equivalent to their product being pure imaginary. We can therefore use the dot product to ensure that the product $\widehat{s}\,\widehat{t}$ in (4.32) is pure imaginary: If not, simply replace $\widehat{t}$ by its orthogonal component $\widehat{t} - (\widehat{t} \cdot \widehat{s})\,\widehat{s}$ (rescaled to have norm one).

What about the cross product? The cross product of two vectors in $\mathbb{R}^3$ points in the unique direction (up to sign) orthogonal to the plane spanned

by the given vectors. In higher dimensions, there is in general no such preferred direction; there are many directions perpendicular to a given plane.
It is therefore somewhat surprising that restricting the octonionic product
to imaginary octonions yields a *cross product* in $\mathbb{R}^7$ with the usual properties (which are essentially linearity, anticommutativity, and orthogonality
to each original vector; there is also a condition on the norm of the cross
product). Remarkably, such products exist *only* in $\mathbb{R}^3$ and $\mathbb{R}^7$, corresponding to imaginary quaternions and octonions, respectively.

One way to specify a unique direction in $\mathbb{R}^n$ is to give $n - 1$ directions
orthogonal to it. One might therefore suspect that there is a generalized
"cross product" in $\mathbb{R}^n$ involving $n - 1$ vectors. This is correct; the resulting
product is most easily described in the language of differential forms. But
there is precisely one further "generalized cross product" of more than two
vectors, namely a product of three vectors in $\mathbb{R}^8$, which also turns out to
be related to the octonions. This *triple cross product* is defined by

$$x \times y \times z = \frac{1}{2}(x(\overline{y}z) - z(\overline{y}x)) \tag{4.39}$$

for any octonions x, y, z. (Note that this is not an iterated cross product,
but a product directly defined on three factors.) The real part of this
product defines the *associative 3-form* Φ, namely

$$\Phi(x, y, z) = \mathrm{Re}\,(x \times y \times z) = \frac{1}{2}\,\mathrm{Re}\,([x, \overline{y}]z) \tag{4.40}$$

which we will use later. (The last equality follows by direct computation.)

As with quaternions, a useful strategy for solving problems is to break
up the octonions into complex or quaternionic pieces. We have

$$x = x_{1\mathbb{H}} + x_{2\mathbb{H}}\ell$$
$$= x_{1\mathbb{C}} + x_{2\mathbb{C}}i + x_{3\mathbb{C}}j + x_{4\mathbb{C}}k \tag{4.41}$$

where $x_{1\mathbb{H}}, x_{2\mathbb{H}} \in \mathbb{H}$ and $x_{1\mathbb{C}}, x_{2\mathbb{C}}, x_{3\mathbb{C}}, x_{4\mathbb{C}} \in \mathbb{C}$, and where our default conventions are that $\mathbb{H}$ is the quaternionic subalgebra generated by i, j, k, but $\mathbb{C}$
is the complex subalgebra determined by ℓ.

- *What is the result of conjugating an octonion by ℓ?*

First of all, there are no associativity issues here since there are only
two octonions involved, namely ℓ and x. The easiest way to work out this
result is to expand x as in (4.2), noting that ℓ anticommutes with each term
except the first and last, so that

$$\ell y \overline{\ell} = y_1 - y_2 i - y_3 j - y_4 k - y_5 k\ell - y_6 j\ell - y_7 i\ell + y_8\ell. \tag{4.42}$$

As a special case, for $q \in \mathbb{H}$ we have

$$\ell q \overline{\ell} = \overline{q} \tag{4.43}$$

and similarly

$$\ell(q\ell)\overline{\ell} = \overline{q}\ell \tag{4.44}$$

or equivalently

$$\ell q = \overline{q}\ell, \tag{4.45}$$

$$\ell(q\ell) = -\overline{q}. \tag{4.46}$$

These last two expressions are pieces of the general multiplication rule, which can be expressed in the form

$$(x_{1\mathbb{H}} + x_{2\mathbb{H}}\ell)(y_{1\mathbb{H}} + y_{2\mathbb{H}}\ell) = (x_{1\mathbb{H}}y_{1\mathbb{H}} - \overline{y_{2\mathbb{H}}}x_{2\mathbb{H}}) + (y_{2\mathbb{H}}x_{1\mathbb{H}} + x_{2\mathbb{H}}\overline{y_{1\mathbb{H}}})\ell \tag{4.47}$$

and from which (4.43) and (4.44) could have been derived, in the form

$$\ell(y_{1\mathbb{H}} + y_{2\mathbb{H}}\ell)\overline{\ell} = \overline{y_{1\mathbb{H}}} + \overline{y_{2\mathbb{H}}}\ell. \tag{4.48}$$

- *What is the result of conjugating an octonion by $e^{\ell\theta}$?*

This follows immediately from the similar computation (3.30) over the quaternions. Write x in terms of four complex numbers as above. Looking first at the quaternionic subalgebra containing i and ℓ, and then replacing i in turn with j and k, leads us to

$$e^{\ell\theta}xe^{-\ell\theta} = e^{\ell\theta}x_{1\mathrm{C}}e^{-\ell\theta} + e^{\ell\theta}x_{2\mathrm{C}}ie^{-\ell\theta} + e^{\ell\theta}x_{3\mathrm{C}}je^{-\ell\theta} + e^{\ell\theta}x_{4\mathrm{C}}ke^{-\ell\theta}$$
$$= x_{1\mathrm{C}} + x_{2\mathrm{C}}e^{2\ell\theta}i + x_{3\mathrm{C}}e^{2\ell\theta}j + x_{4\mathrm{C}}e^{2\ell\theta}k. \tag{4.49}$$

As we will see later, this operation corresponds to a rotation by 2θ in three planes at once!

Chapter 5

Other Number Systems

5.1 The Cayley–Dickson Process

We have constructed the complex numbers, the quaternions, and the octonions by doubling a smaller algebra. We have

$$\mathbb{C} = \mathbb{R} \oplus \mathbb{R}i, \tag{5.1}$$

$$\mathbb{H} = \mathbb{C} \oplus \mathbb{C}j, \tag{5.2}$$

$$\mathbb{O} = \mathbb{H} \oplus \mathbb{H}\ell. \tag{5.3}$$

We can emphasize this doubling, using a slightly different notation. A complex number z is equivalent to a pair of real numbers, its real and imaginary parts. So we can write

$$z = (x, y) \tag{5.4}$$

corresponding in more traditional language to $z = x + iy$. Conjugation and complex multiplication then become

$$\overline{(a, b)} = (a, -b), \tag{5.5}$$

$$(a, b)(c, d) = (ac - bd, bc + ad), \tag{5.6}$$

$$(a, b)\overline{(a, b)} = (a^2 + b^2, 0). \tag{5.7}$$

A quaternion q can be written as a pair of complex numbers,

$$q = (z, w) \tag{5.8}$$

corresponding to $q = z + wj$. Conjugation now takes the form

$$\overline{(a, b)} = (\overline{a}, -b) \tag{5.9}$$

but what about quaternionic multiplication? Working out $(a + bj)(c + dj)$ with $a, b, c, d \in \mathbb{C}$, we see that

$$(a, b)(c, d) = (ac - b\overline{d}, ad + b\overline{c}) \tag{5.10}$$

so that

$$(a, b)\overline{(a, b)} = (|a|^2 + |b|^2, 0). \tag{5.11}$$

Finally, if we write an octonion p as two quaternions (q, r), corresponding to $p = q + r\ell$, we obtain

$$\overline{(a, b)} = (\overline{a}, -b), \tag{5.12}$$

$$(a, b)(c, d) = (ac - \overline{d}b, da + b\overline{c}), \tag{5.13}$$

$$(a, b)\overline{(a, b)} = (|a|^2 + |b|^2, 0). \tag{5.14}$$

All of the above constructions are special cases of the *Cayley–Dickson* process, for which

$$\overline{(a, b)} = (\overline{a}, -b), \tag{5.15}$$

$$(a, b)(c, d) = (ac - \epsilon\overline{d}b, da + b\overline{c}), \tag{5.16}$$

$$(a, b)\overline{(a, b)} = (|a|^2 + \epsilon|b|^2, 0), \tag{5.17}$$

where $\epsilon = \pm 1$. We can use this construction to generate larger algebras from smaller ones, by making successive choices of ϵ at each step.

5.2   Sedenions

What happens if we continue this process? We define the *sedenions* by considering pairs of octonions,

$$s = (p, q) \tag{5.18}$$

with $p, q \in \mathbb{O}$. Sedenion multiplication is defined by the Cayley–Dickson process, with $\epsilon = 1$, so we have

$$\overline{(a, b)} = (\overline{a}, -b), \tag{5.19}$$

$$(a, b)(c, d) = (ac - \overline{d}b, da + b\overline{c}), \tag{5.20}$$

$$(a, b)\overline{(a, b)} = (|a|^2 + |b|^2, 0). \tag{5.21}$$

If we define the special element

$$e = (0, 1) \tag{5.22}$$

then we could also write

$$s = p + qe \tag{5.23}$$

since

$$(p, 0)(1, 0) = (p, 0), \tag{5.24}$$

$$(q, 0)(0, 1) = (0, q). \tag{5.25}$$

The sedenions possess some curious properties. They are of course neither commutative nor associative, since they contain a copy of the octonions. And they do possess a positive-definite inner product; the norm $s\bar{s}$ of any nonzero sedenion s is strictly positive. However, the sedenions contain zero divisors, that is, nonzero elements whose product is nonetheless zero. For example, we have

$$\begin{aligned}
(i\ell + je)(j\ell + ie) &= (i\ell, j)(j\ell, i)\\
&= \big((i\ell)(j\ell) + ij, i(i\ell) - j(j\ell)\big)\\
&= (-k + k, -\ell + \ell) = 0.
\end{aligned} \tag{5.26}$$

Thus, the sedenions are not a composition algebra, as they fail to satisfy the identity

$$|pq| = |p|\,|q| \tag{5.27}$$

and they are not a division algebra, since, for example, zero divisors such as $i\ell + je$ cannot have an inverse.

5.3 The Hurwitz Theorem

A *composition algebra* $\mathbb{K}$ possesses a norm, that is a nondegenerate quadratic form satisfying the identity[1]

$$|pq|^2 = |p|^2|q|^2 \tag{5.28}$$

for all $p, q \in \mathbb{K}$. The *Hurwitz Theorem*, published posthumously by Adolf Hurwitz in 1923, states that the reals, complexes, quaternions, and octonions are the only real composition algebras with positive-definite norm, and hence the only composition algebras without zero divisors. That is, the only such algebras that contain $\mathbb{R}$ are $\mathbb{K} = \mathbb{R}, \mathbb{C}, \mathbb{H}, \mathbb{O}$.

More generally, all real composition algebras can be obtained from the Cayley–Dickson process—and must have dimension 1, 2, 4 or 8. Composition algebras of dimension 1 or 2 are both commutative and associative, composition algebras of dimension 4 are associative but not commutative, and composition algebras of dimension 8 are neither. The proof of the Hurwitz Theorem amounts to showing that the Cayley–Dickson process can only yield a composition algebra starting from an associative algebra.

---

[1] More correctly, the quadratic form acting on an element $q \in \mathbb{K}$ gives the *squared norm* $|q|^2$ of q, which can however be positive, negative, or zero; the *norm* $|q|$ of q makes sense only in the positive-definite case.

The Hurwitz theorem does however leave open the possibility of composition algebras other than $\mathbb{R}, \mathbb{C}, \mathbb{H}, \mathbb{O}$, so long as the norm is *not* positive-definite (or negative-definite, which amounts to the same thing). Furthermore, all such algebras can be constructed using the Cayley–Dickson process, making suitable choices of ϵ at each step. Starting with the reals, which are 1-dimensional, we can apply the Cayley–Dickson process up to three times. Remarkably, however, only three distinct new composition algebras are obtained, the so-called *split* versions of the composition algebras.

5.4   Split Complex Numbers

Start with the real numbers, and apply the Cayley–Dickson process with $\epsilon = -1$. The resulting algebra is known as the *split complex numbers*, denoted $\mathbb{C}'$, and satisfies

$$\mathbb{C}' = \mathbb{R} \oplus \mathbb{R}L \tag{5.29}$$

where

$$L^2 = 1 \tag{5.30}$$

rather than -1. What are the properties of such numbers?

A general element of $\mathbb{C}'$ takes the form $a + bL$, with $a, b \in \mathbb{R}$. Just like the ordinary complex numbers, the split complex numbers are both commutative and associative. The (squared) norm is given by

$$|a + bL|^2 = (a + bL)(a - bL) = a^2 - b^2 \tag{5.31}$$

which is not positive-definite. In particular, $\mathbb{C}'$ contains *zero divisors*, for instance

$$(1 + L)(1 - L) = 1 - L^2 = 0. \tag{5.32}$$

Furthermore

$$\left(\frac{1}{2}(1 \pm L) \right)^2 = \frac{1}{2}(1 \pm L) \tag{5.33}$$

so that $\frac{1}{2}(1 \pm L)$ act as orthogonal projection operators, dividing $\mathbb{C}'$ into two null subspaces.

Another curious property of $\mathbb{C}'$ involves square roots. How many split-complex square roots of unity are there? Four! Not only do ± 1 square to 1, but so do $\pm L$. More generally, from

$$(a + bL)^2 = (a^2 + b^2) + 2abL \tag{5.34}$$

and

$$(a^2 + b^2) \pm 2ab = (a \pm b)^2 \geq 0 \qquad (5.35)$$

we see that a split complex number can only be the square of another split complex number if its real part is at least as large as its imaginary part. In particular, L itself cannot be the square of any split complex number!

You may recognize the inner product (5.31) as that of special relativity in two dimensions, with the spacetime vector (x, t) in 2-dimensional Minkowski space corresponding to the split complex number $x + tL$. The hyperbolic nature of the geometry of special relativity [3] yields

$$e^{L\beta} = \cosh(\beta) + L\sinh(\beta) \qquad (5.36)$$

which can also be checked by expanding $\exp(L\beta)$ as a power series. For this reason, the split complex numbers are also called *hyperbolic numbers*. Hyperbolic numbers not only represent the *points* in 2-dimensional Minkowski space, but can also be used to describe the *Lorentz transformations* between reference frames, which are nothing more than *hyperbolic rotations*.

5.5 Split Quaternions

We can repeat this process to obtain the *split quaternions*, denoted $\mathbb{H}'$, as

$$\mathbb{H}' = \mathbb{C} \oplus \mathbb{C}L. \qquad (5.37)$$

In order to keep track of these different algebras, we will use K rather than i for the imaginary unit here. Thus, $\mathbb{H}'$ consists of linear combinations of 1, K, L, and KL, and it remains to work out the full multiplication table. We have

$$K^2 = -1, \qquad L^2 = +1, \qquad (5.38)$$

$$(K)(L) = KL = -(L)(K), \qquad (5.39)$$

$$(KL)^2 = KLKL = -KKLL = +1, \qquad (5.40)$$

$$K(KL) = -L = -(KL)K, \qquad (5.41)$$

$$(KL)L = K = -L(KL). \qquad (5.42)$$

The split quaternions are associative, but not commutative.

A typical element $Q \in \mathbb{H}'$ thus takes the form

$$Q = Q_1 + Q_2 K + Q_3 KL + Q_4 L \qquad (5.43)$$

and has (squared) norm

$$|Q|^2 = Q\overline{Q} = Q_1^2 + Q_2^2 - Q_3^2 - Q_4^2 \qquad (5.44)$$

Table 5.1   The split octonionic multiplication table.

	I	J	K	KL	JL	IL	L
I	-1	K	$-J$	JL	$-KL$	$-L$	IL
J	$-K$	-1	I	$-IL$	$-L$	KL	JL
K	J	$-I$	-1	$-L$	IL	$-JL$	KL
KL	$-JL$	IL	L	1	$-I$	J	K
JL	KL	L	$-IL$	I	1	$-K$	J
IL	L	$-KL$	JL	$-J$	K	1	I
L	$-IL$	$-JL$	$-KL$	$-K$	$-J$	$-I$	1

which has *signature* $(2,2)$, that is, two of our basis elements have (squared) norm $+1$, namely 1 and K, and two have (squared) norm -1, namely KL and L.

The split quaternions can also be obtained as

$$\mathbb{H}' = \mathbb{C}' \oplus \mathbb{C}'K \tag{5.45}$$

so there are only two 4-dimensional composition algebras over the reals, namely $\mathbb{H}$ and $\mathbb{H}'$, not three. That is, it doesn't matter whether the Cayley–Dickson process is done first with $\epsilon = +1$, then with $\epsilon = -1$, or vice versa. Note, however, that the split quaternions $\mathbb{H}'$ contain *both* the split complex numbers $\mathbb{C}'$ and the ordinary complex numbers $\mathbb{C}$.

5.6   Split Octonions

If we repeat this process one more time, we obtain the *split octonions*, denoted $\mathbb{O}'$, as

$$\mathbb{O}' = \mathbb{H} \oplus \mathbb{H}L \tag{5.46}$$

where we now use I, J, K for the imaginary units in $\mathbb{H}$. Thus, $\mathbb{O}'$ consists of linear combinations of $\{1, I, J, K, KL, JL, IL, L\}$, and it again remains to work out the full multiplication table; the result is shown in Table 5.1. The split octonions are not associative, but they are alternative.

It is easily checked that the inner product on $\mathbb{O}'$ has signature $(4,4)$; our conventions are such that basis elements containing L have (squared) norm -1, and all others have (squared) norm $+1$.

The split octonions can also be obtained as

$$\mathbb{O}' = \mathbb{H}' \oplus \mathbb{H}'J \tag{5.47}$$

so there are again only two 8-dimensional composition algebras over the reals, namely $\mathbb{O}$ and $\mathbb{O}'$. As before, the split octonions $\mathbb{O}'$ contain both the split quaternions $\mathbb{H}'$ and the ordinary quaternions $\mathbb{H}$.

5.7 Subalgebras of the Split Octonions

Unlike the (ordinary) octonions, the split octonions have subalgebras that are not themselves composition algebras. The composition property itself doesn't fail; if it holds in the full algebra, it holds in all subalgebras. Rather, it is the requirement that the norm be nondegenerate that fails; these subalgebras are all null (to various degrees).

We have already seen that $\frac{1}{2}(1 \pm L)$ are projection operators in $\mathbb{C}'$. This means that the subalgebra $\langle 1 + L \rangle$ (consisting of all real multiples of $1 + L$) closes under multiplication, and is therefore a subalgebra of $\mathbb{C}' \subset \mathbb{O}'$. This subalgebra is *isomorphic* to the real numbers, since $\frac{1}{2}(1 + L)$ acts as an identity element, but is not *isometric* to the real numbers, since $|1 + L| = 0$. All elements of this subalgebra are null, that is, have norm zero.

Other null elements of $\mathbb{O}'$ also generate 1-dimensional null subalgebras, such as $\langle I + IL \rangle$. In this case, not only is $|I + IL| = 0$, but also $(I + IL)^2 = 0$; all products in this subalgebra are zero.

We can combine such null subalgebras in various ways. The 2-dimensional subalgebra $\langle I + IL, J - JL \rangle$ again has all products zero, whereas the 2-dimensional subalgebra $\langle 1 + L, I + IL \rangle$ does not. This latter algebra has the peculiar property that there are elements whose product is zero in one order, but not the other, since

$$(1 + L)(I + IL) = 0, \tag{5.48}$$

$$(I + IL)(1 + L) = 2(I + IL). \tag{5.49}$$

Similarly, there are 3-dimensional null subalgebras $\langle I + IL, J + JL, K - KL \rangle$ and $\langle 1 + L, I + IL, J - JL \rangle$, as well as a 4-dimensional null subalgebra $\langle 1 + L, I + IL, J + JL, K - KL \rangle$.

Each of the subalgebras above is *totally null*; every element has norm zero. There are also subalgebras of $\mathbb{O}'$ that are only partially null; each such subalgebra contains the identity element 1.

The best-known of these subalgebras are the 3-dimensional *ternions*, generated by $\{1, L, I + IL\}$, and the 6-dimensional *sextonions*, generated by $\{1, I, IL, L, J + JL, K - KL\}$; there is also a 4-dimensional subalgebra generated by $\{1, L, I + IL, K - KL\}$.

PART II
Symmetry Groups

Chapter 6

Some Orthogonal Groups

6.1 Rotations

Rotations in two dimensions are easily described; just specify the angle of rotation and the orientation (clockwise or counterclockwise). In three dimensions, it is also necessary to specify the axis of rotation. By convention, a positive angle of rotation about that axis implies rotation in a *counterclockwise* direction looking back along the axis. Thus, a rotation by $\frac{\pi}{2}$ about the North Pole corresponds to spinning the globe $\frac{1}{4}$ of the way around to the *east*; spinning to the west would correspond to a negative rotation about the North Pole, or equivalently to a positive rotation about the South Pole.

How do you specify a rotation in higher dimensions? In three dimensions, specifying the axis of rotation is just a way of specifying the plane in which the rotation takes place. In higher dimensions, one must specify the plane of rotation itself, as there is more than one "axis" perpendicular to any plane.

A key property of rotations is that they preserve length! Rotations take a sphere to itself, and hence take any vector to some other vector of the same length. In other words, the norm $|\vec{v}|$ of a vector is unchanged by a rotation. However, rotations are not the only transformations with this property; there are also reflections. The difference is that rotations are *orientation-preserving*, whereas reflections are *orientation-reversing*. (These are the only possibilities; any linear transformation must result in either an even or an odd permutation of the relative positions of the axes.)

So in higher dimensions, we *define* a rotation to be a length-preserving, orientation-preserving linear transformation of a real vector space. If you preserve length and orientation, and then do so again, you have clearly preserved length and orientation, so the composition of two rotations is again a rotation. Furthermore, you can undo any (sequence of) rotation(s) simply by performing the same rotation(s) in the opposite direction (and in the opposite order). These two properties—closure under composition and the existence of an inverse—ensure that the collection of all rotations is a *group*.[1]

The group of length-preserving linear transformations in n dimensions is called O(n), the *orthogonal group in n dimensions*; those transformations which are also orientation-preserving make up the *rotation group in n dimensions*, SO(n).

These groups are normally expressed in terms of matrices. Since we are studying transformations of $\mathbb{R}^n$, these matrices are real, so that matrix multiplication is associative. Consider first linear transformations in two dimensions, which take a vector

$$v = \begin{pmatrix} x \\ y \end{pmatrix} \tag{6.1}$$

to another vector

$$w = Mv \tag{6.2}$$

where M is some matrix

$$M = \begin{pmatrix} a & b \\ c & d \end{pmatrix} \tag{6.3}$$

with $a, b, c, d, x, y \in \mathbb{R}$. The (squared) length of v is given by

$$|v|^2 = x^2 + y^2 = v^T v \tag{6.4}$$

where v^T denotes the matrix transpose of v. Thus, the *orthogonal group* O(2) consists of those matrices M that preserve the length of v, that is, for which

$$v^T v = (Mv)^T (Mv) = v^T M^T M v \tag{6.5}$$

for any $v \in \mathbb{R}^2$. This can only be true if

$$M^T M = I \tag{6.6}$$

where I is the (2×2) identity matrix. Since

$$\det(M^T) = \det(M) \tag{6.7}$$

---

[1]The group operation must also be *associative*. But composition of transformations is associative by definition.

we must have

$$\det M = \pm 1 \tag{6.8}$$

and it is straightforward to generalize this construction to higher dimensions. The rotation groups can therefore be expressed as

$$O(n) = \{M \in \mathbb{R}^{n \times n} : M^T M = I\}, \tag{6.9}$$

$$SO(n) = \{M \in \mathbb{R}^{n \times n} : M^T M = I, \det M = 1\}, \tag{6.10}$$

where we have introduced the notation $\mathbb{K}^{m \times n}$ for the $m \times n$ matrices whose elements lie in $\mathbb{K}$; these expressions are often taken as the *definitions* of the *orthogonal* and *special orthogonal* groups, respectively.

In the remainder of this chapter, we set the stage for later developments by discussing the properties of several important orthogonal groups. In certain dimensions, orthogonal transformations can also be expressed in terms of division algebras other than $\mathbb{R}$; we save that discussion for Sections 7.2 and 9.1.

6.2 The Geometry of SO(2)

From the definition in the previous section, we have

$$SO(2) = \{M \in \mathbb{R}^{2 \times 2} : M^T M = I, \det M = 1\}. \tag{6.11}$$

It is easy to show that the most general element of $SO(2)$ takes the form

$$M = \begin{pmatrix} \cos \alpha & -\sin \alpha \\ \sin \alpha & \cos \alpha \end{pmatrix} \tag{6.12}$$

representing a counterclockwise rotation by α in the xy plane. It is in fact enough to check this claim on a basis, such as

$$\hat{\imath} = \begin{pmatrix} 1 \\ 0 \end{pmatrix}, \tag{6.13}$$

$$\hat{\jmath} = \begin{pmatrix} 0 \\ 1 \end{pmatrix}, \tag{6.14}$$

noting that

$$M\hat{\imath} = \begin{pmatrix} \cos \alpha \\ \sin \alpha \end{pmatrix}, \tag{6.15}$$

$$M\hat{\jmath} = \begin{pmatrix} -\sin \alpha \\ \cos \alpha \end{pmatrix}, \tag{6.16}$$

which indeed correspond to a counterclockwise rotation of the x and y axes through an angle α, resulting in the rotated basis $\{M\hat{\imath}, M\hat{\jmath}\}$—which are just the columns of M.

6.3   The Geometry of SO(3)

In three dimensions, every rotation is in fact a rotation about some single axis. To specify a rotation in three dimensions, it is therefore necessary to specify this axis, and the angle about this axis through which to rotate. It takes two parameters, such as latitude and longitude, to determine the location of the axis, and a third to give the angle of rotation. These three parameters are collectively known as *Euler angles.*

Another way to describe rotations in three dimensions is to explicitly construct rotations (only) in the coordinate planes, and then argue that any rotation can be obtained by suitably combining these rotations. In particular, it is well-known that any rotation matrix in three dimensions can be written (in several ways) as the product of three such rotations in coordinate planes; this is in fact one way to construct the Euler angle representation to begin with.

Since a description in terms of generalized Euler angles is not available in higher dimensions, we focus instead on the alternative description in terms of rotations in coordinate planes.

Rotation matrices of the form (6.12) generalize in an obvious way to higher dimensions: Just make the matrix bigger, and add 1 and 0 appropriately in the remaining entries. This construction leads to the three matrices

$$R_x = R_x(\alpha) = \begin{pmatrix} 1 & 0 & 0 \\ 0 & \cos\alpha & -\sin\alpha \\ 0 & \sin\alpha & \cos\alpha \end{pmatrix}, \tag{6.17}$$

$$R_y = R_y(\alpha) = \begin{pmatrix} \sin\alpha & 0 & \cos\alpha \\ 0 & 1 & 0 \\ \cos\alpha & 0 & -\sin\alpha \end{pmatrix}, \tag{6.18}$$

$$R_z = R_z(\alpha) = \begin{pmatrix} \cos\alpha & -\sin\alpha & 0 \\ \sin\alpha & \cos\alpha & 0 \\ 0 & 0 & 1 \end{pmatrix}, \tag{6.19}$$

corresponding to rotations in the yz, zx, and xy planes, respectively.

Thus, the orthogonal group in three dimensions is *defined* by

$$\text{SO}(3) = \{M \in \mathbb{R}^{3\times3} : M^T M = I, \det M = 1\} \tag{6.20}$$

but is *generated* by the matrices R_x, R_y, R_z. We write this relationship as either of

$$\text{SO}(3) = \langle\{R_x, R_y, R_z\}\rangle = \langle R_x, R_y, R_z \rangle \tag{6.21}$$

where care must be taken to avoid confusing this language with the similar notation used to denote the span of certain vectors in a vector space. We interpret (6.21) as meaning that any matrix in SO(3) can be written as the product of matrices of the form R_x, R_y, R_z; there is no restriction on how many matrices might be needed, nor on what parameter values (rotation angles) are allowed.

6.4 The Geometry of SO(4)

In higher dimensions, not every rotation corresponds to a single axis together with an angle of rotation about this axis; the concept of Euler angles does not generalize. The first problem is that, even for rotations in a given plane, it is not possible to associate an "axis" with a given rotation, as there are multiple directions orthogonal to the plane of rotation. Furthermore, it is no longer the case that every rotation corresponds to rotation in a single plane.

We first encounter this difficulty in four dimensions, where we can rotate each of two independent planes arbitrarily. For this reason, we make no effort to identify all matrices satisfying the definition

$$SO(4) = \{M \in \mathbb{R}^{4 \times 4} : M^T M = I, \det M = 1\} \tag{6.22}$$

but rather rely on the fact that any such rotation can be generated by rotations in coordinate planes. How many such planes are there? In n dimensions, there are $\binom{n}{2}$ possible planes, so the dimension of the orthogonal groups is given by

$$|SO(n)| = \frac{1}{2} n(n - 1). \tag{6.23}$$

In four dimensions, there are $\binom{4}{2} = 6$ possible planes; there are six matrices of the same general form as our R_x, R_y, R_z from SO(3). In SO(3), however, we labeled our generators by the axis about which they rotate; in higher dimensions, we must instead label them by the plane in which they rotate. So start by relabeling our generators of SO(3) as R_{yz}, R_{zx}, R_{yz}, and reinterpret them as transformations in four dimensions that hold the fourth axis, w say, fixed. Then the three remaining generators of SO(4) are R_{wx}, R_{wy}, R_{wz}, and we have

$$SO(4) = \langle R_{yz}, R_{zx}, R_{yz}, R_{wx}, R_{wy}, R_{wz} \rangle \tag{6.24}$$

Each of these generators of SO(4) corresponds to a rotation in a single plane, leaving the orthogonal plane invariant. We can instead consider rotations that rotate two orthogonal planes. Although the angles of rotation in the two planes could be different, we consider the special case where these angles are equal in magnitude. Such rotations are called *isoclinic*. Consider therefore the transformations

$$S_{yz}^{\pm} = S_{yz}^{\pm}(\alpha) = R_{yz}(\alpha)R_{wx}(\pm\alpha), \qquad (6.25)$$

$$S_{zx}^{\pm} = S_{zx}^{\pm}(\alpha) = R_{zx}(\alpha)R_{wy}(\pm\alpha), \qquad (6.26)$$

$$S_{xy}^{\pm} = S_{xy}^{\pm}(\alpha) = R_{xy}(\alpha)R_{wz}(\pm\alpha), \qquad (6.27)$$

and the subsets

$$\mathrm{SO4}^{\pm} = \langle S_{yz}^{\pm}, S_{zx}^{\pm}, S_{xy}^{\pm} \rangle \qquad (6.28)$$

of SO(4). Remarkably, each of these subsets closes under multiplication; SO4$^{\pm}$ are *subgroups* of SO(4). It is not hard to see that the multiplication table for the generators of SO4$^{\pm}$ is identical to that of SO(3); we say that

$$\mathrm{SO4}^{\pm} \cong \mathrm{SO}(3) \qquad (6.29)$$

("each of SO4$^{\pm}$ is *isomorphic* to SO(3)"). Furthermore, we can recover all of SO(4) by multiplying elements of these two subgroups together. For instance,

$$S_{xy}^{+}(\alpha)S_{xy}^{-}(\alpha) = R_{xy}(2\alpha), \qquad (6.30)$$

$$S_{xy}^{+}(\alpha)S_{xy}^{-}(-\alpha) = R_{wx}(2\alpha), \qquad (6.31)$$

and we have shown that

$$\mathrm{SO}(3) \times \mathrm{SO}(3) = \mathrm{SO}(4) \qquad (6.32)$$

a result that we will revisit in quaternionic language in Section 9.1.2.

6.5 Lorentz Transformations

There is another kind of "rotation", namely the Lorentz transformations of special relativity that relate inertial reference frames. Geometrically, such transformations preserve a generalized distance, the (squared) "interval" between spacetime events.

So in two dimensions, consider the vector

$$v = \begin{pmatrix} t \\ x \end{pmatrix} \qquad (6.33)$$

which represents a spacetime event occurring at time t and position x.[2] The (squared) magnitude of v is defined by

$$|v|^2 = x^2 - t^2 \tag{6.34}$$

and represents the (squared) *spacetime interval* between the point (t, x) and the origin. We can rewrite (6.34) in matrix language by introducing the *metric* g, which here takes the form

$$g = \begin{pmatrix} -1 & 0 \\ 0 & 1 \end{pmatrix} \tag{6.35}$$

leading to

$$|v|^2 = v^T g v. \tag{6.36}$$

The *signature* of an invertible diagonal matrix such as g is the number of positive and negative entries; the signature of the identity matrix I (in two dimensions) is $(2, 0)$, and the signature of g is $(1, 1)$.

As with ordinary rotations, we can now seek those linear transformations that preserve the magnitude of v, that is, for which

$$v^T v = (Mv)^T g(Mv) = v^T M^T g M v \tag{6.37}$$

for any $v \in \mathbb{R}^2$, which can only be true if

$$M^T g M = g. \tag{6.38}$$

As with ordinary rotations, we must have $\det M = \pm 1$; we will consider only the case where $\det M = +1$. We can therefore define the *generalized rotation groups* $SO(p, q)$ by

$$SO(p, q) = \{M \in \mathbb{R}^{n \times n} : M^T g M = g, \det M = 1\} \tag{6.39}$$

where g now has signature (p, q). If $q = 0$ (or $p = 0$), we recover the ordinary rotation groups; if $q = 1$, corresponding to a single "timelike" direction, we obtain the *Lorentz group in $p + 1$ dimensions*. One often writes $\mathbb{R}^{p,q}$ for the vector space with metric g of signature (p, q), and $\mathbb{M}^{p+1} = \mathbb{R}^{p,1}$ for *Minkowski space in $p + 1$ dimensions*, the arena for special relativity.

The geometry of 2-dimensional Minkowski space is particularly interesting, and is discussed in more detail in [3]. From the definition above, we have

$$SO(1, 1) = \{M \in \mathbb{R}^{2 \times 2} : M^T g M = g, \det M = 1\}. \tag{6.40}$$

---

[2] We adopt units in which the speed of light $c = 1$, effectively replacing t by ct, and thus measuring time in units of length, such as centimeters.

It is easy to show that the most general element of $SO(1,1)$ takes the form

$$M = \begin{pmatrix} \cosh\alpha & \sinh\alpha \\ \sinh\alpha & \cosh\alpha \end{pmatrix} \tag{6.41}$$

representing a *boost* in the x direction, that is, a Lorentz transformation from the laboratory reference frame, at rest, to a reference frame moving to the right with speed $\tanh\alpha$.

6.6   The Geometry of $SO(3,1)$

The world around us appears to have three independent directions, namely East/West, North/South, and up/down. Special relativity tells us to include time; now we have a fourth "direction", namely toward the future or past.

It is customary to label these directions with time first, so introduce coordinates $\{t, x, y, z\}$. A spacetime vector is therefore a vector with *four* components, such as

$$v = \begin{pmatrix} t \\ x \\ y \\ z \end{pmatrix} \tag{6.42}$$

which is also called a 4-vector.

The inner product on 4-dimensional Minkowski space, the arena for special relativity, is given by (6.36), where now

$$g = \begin{pmatrix} -1 & 0 & 0 & 0 \\ 0 & 1 & 0 & 0 \\ 0 & 0 & 1 & 0 \\ 0 & 0 & 0 & 1 \end{pmatrix} \tag{6.43}$$

so that

$$|v|^2 = v^T g v = -t^2 + x^2 + y^2 + z^2 \tag{6.44}$$

which is also called the *squared interval* between the *event* (t, x, y, z) and the origin.

The *Lorentz group* in $3+1$ spacetime dimensions is given by

$$SO(3,1) = \{M \in \mathbb{R}^{4\times4} : M^T g M = g, \det M = 1\}. \tag{6.45}$$

As with SO(4), it is sufficient to give generators for SO(3, 1), namely generalized rotations in the $\binom{4}{2} = 6$ independent coordinate planes. As before, we have the *rotations*

$$R_{yz} = R_{yz}(\alpha) = \begin{pmatrix} 1 & 0 & 0 & 0 \\ 0 & 1 & 0 & 0 \\ 0 & 0 & \cos\alpha & -\sin\alpha \\ 0 & 0 & \sin\alpha & \cos\alpha \end{pmatrix}, \tag{6.46}$$

$$R_{zx} = R_{zx}(\alpha) = \begin{pmatrix} 1 & 0 & 0 & 0 \\ 0 & \sin\alpha & 0 & \cos\alpha \\ 0 & 0 & 1 & 0 \\ 0 & \cos\alpha & 0 & -\sin\alpha \end{pmatrix}, \tag{6.47}$$

$$R_{xy} = R_{xy}(\alpha) = \begin{pmatrix} 1 & 0 & 0 & 0 \\ 0 & \cos\alpha & -\sin\alpha & 0 \\ 0 & \sin\alpha & \cos\alpha & 0 \\ 0 & 0 & 0 & 1 \end{pmatrix}, \tag{6.48}$$

corresponding to rotations in the yz, zx, and xy planes, respectively. But we also have the *Lorentz transformations* given by

$$R_{tx} = R_{tx}(\alpha) = \begin{pmatrix} \cosh\alpha & \sinh\alpha & 0 & 0 \\ \sinh\alpha & \cosh\alpha & 0 & 0 \\ 0 & 0 & 1 & 0 \\ 0 & 0 & 0 & 1 \end{pmatrix}, \tag{6.49}$$

$$R_{ty} = R_{ty}(\alpha) = \begin{pmatrix} \cosh\alpha & 0 & \sinh\alpha & 0 \\ 0 & 1 & 0 & 0 \\ \sinh\alpha & 0 & \cosh\alpha & 0 \\ 0 & 0 & 0 & 1 \end{pmatrix}, \tag{6.50}$$

$$R_{tz} = R_{tz}(\alpha) = \begin{pmatrix} \cosh\alpha & 0 & 0 & \sinh\alpha \\ 0 & 1 & 0 & 0 \\ 0 & 0 & 1 & 0 \\ \sinh\alpha & 0 & 0 & \cosh\alpha \end{pmatrix}, \tag{6.51}$$

corresponding to *hyperbolic rotations* in the tx, ty, and tz planes, respectively, which are also called *boosts* in the x, y, and z directions, respectively. Any element of the Lorentz group can be obtained as a product of these six generators (with suitable parameters).[3]

---

[3]These six transformations actually generate the *restricted Lorentz group*, as we have excluded transformations that reverse both the time direction and the spatial orientation.

As was the case in SO(4), not every element of SO(3, 1) corresponds to a rotation or boost in a single plane. But there are also elements of SO(3, 1) that do not correspond to rotations or boosts in any plane.

Both the rotation R_{xy} and the boosts R_{tx} and R_{ty} leave the z direction fixed; this rotation also fixes a *timelike* direction (t), whereas these boosts also fix a *spacelike* direction (y and x, respectively). But consider the matrix

$$N = N(\alpha) = \begin{pmatrix} 1 + \alpha^2/2 & \alpha & -\alpha^2/2 & 0 \\ \alpha & 1 & -\alpha & 0 \\ \alpha^2/2 & \alpha & 1 - \alpha^2/2 & 0 \\ 0 & 0 & 0 & 1 \end{pmatrix} \in \text{SO}(3, 1) \qquad (6.52)$$

which also fixes the z direction. What else does this transformation leave fixed? Direct computation shows that N fixes the *null* direction $y = t$; N is a *null rotation*.

6.7   The Geometry of SO(4, 2)

The Lorentz group preserves "lengths", that is, it preserves the squared interval. We can ask instead what transformations preserve "angles".

In Euclidean space, we can define angles in terms of the dot product, that is, the angle θ between two vectors $\vec{v}$ and $\vec{w}$ is given by

$$\cos\theta = \frac{\vec{v} \cdot \vec{w}}{|\vec{v}|\,|\vec{w}|}. \qquad (6.53)$$

Any transformation that preserves length will preserve angles. But if we rescale $\vec{v}$ and $\vec{w}$, the angle between them still doesn't change.

This formula for the angle between two vectors relies on the Euclidean signature, but we can use the notion of scale-invariance to generalize the notion of "angle-preserving" transformations. A *conformal transformation* is one which preserves the inner product up to scale. Clearly, all orthogonal transformations are also conformal transformations. And if we multiply all vectors by an arbitrary factor, that is if we "stretch" or "dilate" our space by a given amount, then we have simply multiplied all inner products by a corresponding factor. That is,

$$(\lambda v)^T (\lambda w) = \lambda^2 (v^T w) \qquad (6.54)$$

The transformation $v \longmapsto \lambda v$ is called a *dilation*; dilations are another type of conformal transformations.

Conformal transformations should be thought of as acting on the *end-points* of two vectors. If we shift every point in space by the same amount in a given direction, inner products don't change at all; *translations* are also conformal transformations.

There is one additional type of conformal transformation, known as a *conformal translation*. Let $v \in V$ be non-null, so that $|v| \neq 0$. We can then define the inverse of v via

$$v^{-1} = \frac{v}{|v|^2} \tag{6.55}$$

since the inner product of v with v^{-1} is 1. Taking the inverse of a vector "inverts" it through the unit circle, and we use this to define a conformal translation by a constant vector a to be the transformation

$$v \longmapsto \left(v^{-1} + a\right)^{-1} = \frac{v + \alpha|v|^2}{1 + 2\langle v, \alpha \rangle + |\alpha|^2|v|^2}. \tag{6.56}$$

If you see a resemblance between this transformation and the null translations in the previous section, that's not a coincidence.

Conformal translations are clearly not linear. Nonetheless, conformal transformations can be identified with orthogonal transformations. We consider here only the conformal group of Minkowski space, that is, we start with $\mathrm{SO}(3,1)$ and obtain $\mathrm{SO}(4,2)$, but a similar construction can be used in other cases.

Consider the vector

$$V = \begin{pmatrix} T \\ X \\ Y \\ Z \\ P \\ Q \end{pmatrix} \tag{6.57}$$

and define

$$v = \begin{pmatrix} t \\ x \\ y \\ z \end{pmatrix} = \frac{1}{P+Q} \begin{pmatrix} T \\ X \\ Y \\ Z \end{pmatrix}. \tag{6.58}$$

Then we claim that $\mathrm{SO}(4,2)$ acting as usual on V induces conformal transformations acting on v.

We assume that T and Q are the timelike coordinates, that is, that

$$|V|^2 = -T^2 + X^2 + Y^2 + Z^2 + P^2 - Q^2 \tag{6.59}$$

and we further assume that V (but not v!) is null, that is, that

$$|V|^2 = 0. \tag{6.60}$$

Then $SO(3,1) \subset SO(4,2)$ acts as usual on all of T, X, Y, Z, but leaves P and Q alone. This means that $SO(3,1)$ acts as usual on v, since the factor $P + Q$ goes along for the ride. Thus, Lorentz transformations on v are contained in $SO(4,2)$.

Consider now the boost $R_{PQ} \in SO(4,2)$, which leaves T, X, Y, and Z alone, but takes $P + Q$ to

$$(P \cosh\alpha + Q \sinh\alpha) + (Q \cosh\alpha + P \sinh\alpha) = (P + Q) e^{\alpha}. \tag{6.61}$$

Thus, $P + Q$ goes to a multiple of itself, and therefore so does v; this is the dilation on v.

We have accounted for seven of the $\binom{6}{2} = 15$ elements of $SO(4,2)$; the eight remaining transformations are best understood in terms of null rotations. Consider for example

$$R_{\pm} = \begin{pmatrix} 1 & 0 & 0 & 0 & 0 & 0 \\ 0 & 1 & 0 & 0 & \pm\alpha & \alpha \\ 0 & 0 & 1 & 0 & 0 & 0 \\ 0 & 0 & 0 & 1 & 0 & 0 \\ 0 & \mp\alpha & 0 & 0 & 1 - \alpha^2/2 & \mp\alpha^2/2 \\ 0 & \alpha & 0 & 0 & \pm\alpha^2/2 & 1 + \alpha^2/2 \end{pmatrix}. \tag{6.62}$$

Direct computation shows that R_+ induces a translation, and R_- a conformal translation, both in the x direction; similar constructions yield the remaining translations and conformal translations.

We have shown that $SO(4,2)$, the orthogonal group acting on V, is also the conformal group when acting on v, that is, the conformal group of $(3 + 1)$-dimensional Minkowski space is precisely $SO(4,2)$.

Chapter 7

Some Unitary Groups

7.1 Unitary Transformations

Unitary transformations are analogous to rotations, but over the complex numbers, rather than the reals.

We start again in two dimensions. Let v be a *complex* vector

$$v = \begin{pmatrix} w \\ z \end{pmatrix} \tag{7.1}$$

that is, a vector with complex components $w, z \in \mathbb{C}$; we write this as $v \in \mathbb{C}^2$. The squared length of v is given by

$$|v|^2 = |w|^2 + |z|^2 = \overline{w}w + \overline{z}z = v^\dagger v \tag{7.2}$$

where $v^\dagger$ denotes the *Hermitian conjugate* of v, defined by

$$v^\dagger = \begin{pmatrix} \overline{w} & \overline{z} \end{pmatrix} \tag{7.3}$$

or in other words the complex conjugate of the transpose of v. The *unitary group* U(2) consists of those complex matrices M that preserve the length of v, that is, for which

$$v^\dagger v = (Mv)^\dagger (Mv) = v^\dagger M^\dagger M v \tag{7.4}$$

for any $v \in \mathbb{C}^2$, which can only be true if

$$M^\dagger M = I. \tag{7.5}$$

Since

$$\det(M^\dagger) = \overline{\det(M)} \tag{7.6}$$

we must have

$$|\det M| = 1 \tag{7.7}$$

45

and it is straightforward to generalize this construction to higher dimensions. The unitary groups can therefore be expressed as

$$\mathrm{U}(n) = \{M \in \mathbb{C}^{n \times n} : M^\dagger M = I\}, \tag{7.8}$$

$$\mathrm{SU}(n) = \{M \in \mathbb{C}^{n \times n} : M^\dagger M = I, \det M = 1\}, \tag{7.9}$$

which are often taken as the definitions of the *unitary* and *special unitary* groups, respectively. The dimension of the unitary groups is given by

$$|\mathrm{SU}(n)| = n^2 - 1. \tag{7.10}$$

In the remainder of this chapter, we set the stage for later developments by discussing the properties of several important unitary groups. In certain dimensions, unitary transformations can also be expressed in terms of division algebras other than $\mathbb{C}$; we save that discussion for Section 9.2.

7.2   The Geometry of U(1)

We saw in Section 2.4 that complex multiplication can be interpreted geometrically as a rescaling and a rotation. A pure rotation is therefore obtained by multiplying by a *unit* complex number. In other words, if $|w| = 1$, then $|wz| = |z|$, that is, the length of z is preserved under multiplication by w. What do unit-normed elements $w \in \mathbb{C}$ look like? Since $|re^{i\theta}| = r$, we have

$$w = e^{i\theta} \tag{7.11}$$

for some θ. Thus,

$$\mathrm{U}(1) = \{w \in \mathbb{C} : \overline{w}w = 1\} = \{e^{i\theta} : \theta \in [0, 2\pi)\} \tag{7.12}$$

which can also be written as

$$\mathrm{U}(1) = \{w \in \mathbb{C} : |w| = 1\}. \tag{7.13}$$

Equivalently, we can describe U(1) as the group of *transformations*

$$\mathrm{U}(1) = \{z \longmapsto wz : w, z \in \mathbb{C}, |w| = 1\}. \tag{7.14}$$

As already noted, a special case occurs when $\theta = \frac{\pi}{2}$, in which case $w = i$.

But we already have a name for the rotations in the plane, namely SO(2), which we studied in Section 6.2. Thus, SO(2) and U(1) are the same group, which we write as

$$\mathrm{U}(1) \cong \mathrm{SO}(2) \tag{7.15}$$

where the symbol "$\cong$" is read as "is isomorphic to". This isomorphism is just the first of several we will encounter relating different descriptions of the same group.

What about reflections?

A reflection about the y-axis is easy; that's just complex conjugation, namely the map

$$z \longmapsto \overline{z}. \tag{7.16}$$

A reflection about any other line through the origin can be obtained by a combination of rotations and conjugation. For instance, reflection about the line $y = x$ is given by

$$z \longmapsto e^{i\theta/4}\left(\overline{e^{-i\theta/4}z}\right) = i\overline{z} \tag{7.17}$$

and reflection about the x-axis is given by

$$z \longmapsto i\left(\overline{-iz}\right) = -\overline{z}. \tag{7.18}$$

7.3 The Geometry of SU(2)

The unitary group in two complex dimensions is defined by

$$\mathrm{SU}(2) = \{M \in \mathbb{C}^{2\times 2} : M^\dagger M = I, \det M = 1\}. \tag{7.19}$$

A 2×2 matrix has four complex components, the constraint $M^\dagger M = I$ imposes four real conditions, and the determinant restriction adds just one more (since the other conditions already imply that $|\det M| = 1$). Thus, we expect there to be three independent "rotations" in SU(2). A set of generators is given by

$$R_x = R_x(\alpha) = \begin{pmatrix} \cos\left(\frac{\alpha}{2}\right) & -i\sin\left(\frac{\alpha}{2}\right) \\ -i\sin\left(\frac{\alpha}{2}\right) & \cos\left(\frac{\alpha}{2}\right) \end{pmatrix}, \tag{7.20}$$

$$R_y = R_y(\alpha) = \begin{pmatrix} \cos\left(\frac{\alpha}{2}\right) & -\sin\left(\frac{\alpha}{2}\right) \\ \sin\left(\frac{\alpha}{2}\right) & \cos\left(\frac{\alpha}{2}\right) \end{pmatrix}, \tag{7.21}$$

$$R_z = R_z(\alpha) = \begin{pmatrix} e^{-i\alpha/2} & 0 \\ 0 & e^{i\alpha/2} \end{pmatrix}, \tag{7.22}$$

where the factor of 2 is conventional. These matrices are closely related to the *Pauli matrices*

$$\sigma_m = 2i\frac{\partial R_m}{\partial \alpha}\bigg|_{\alpha=0} \tag{7.23}$$

which are explicitly given by

$$\sigma_x = \begin{pmatrix} 0 & 1 \\ 1 & 0 \end{pmatrix},$$ (7.24)

$$\sigma_y = \begin{pmatrix} 0 & -i \\ i & 0 \end{pmatrix},$$ (7.25)

$$\sigma_z = \begin{pmatrix} 1 & 0 \\ 0 & -1 \end{pmatrix}.$$ (7.26)

The Pauli matrices have some important properties: They are *Hermitian* ($\sigma_m^\dagger = \sigma$), tracefree ($\mathrm{tr}\, \sigma_m = 0$), they each square to the identity matrix ($\sigma_m^2 = I$), and they each have determinant -1 ($\det \sigma_m = -1$). We will have more to say about the Pauli matrices later.

How are we to interpret such complex transformations?

A complex vector $v \in \mathbb{C}^2$ can also be viewed as an element of $\mathbb{R}^4$, by treating the real and imaginary parts of the components of v as independent. So consider

$$v = \begin{pmatrix} a \\ b \end{pmatrix} = \begin{pmatrix} a_1 + a_2 i \\ b_1 + b_2 i \end{pmatrix}$$ (7.27)

with $a_1, a_2, b_1, b_2 \in \mathbb{R}$. The matrix

$$\boldsymbol{X} = vv^\dagger = \begin{pmatrix} |a|^2 & a\overline{b} \\ b\overline{a} & |b|^2 \end{pmatrix}$$ (7.28)

has vanishing determinant, since

$$\det \boldsymbol{X} = |a|^2 |b|^2 - |a\overline{b}|^2 = 0.$$ (7.29)

If we further assume that v is normalized, that is, that

$$v^\dagger v = |a|^2 + |b|^2 = 1$$ (7.30)

then we can write

$$\boldsymbol{X} = \begin{pmatrix} \frac{1}{2} + z & x - iy \\ x + iy & \frac{1}{2} - z \end{pmatrix}$$ (7.31)

where

$$x + iy = (a_1 b_1 + a_2 b_2) + (a_1 b_2 - a_2 b_1)i,$$ (7.32)

$$z = \frac{1}{2}(|a|^2 - |b|^2).$$ (7.33)

Since $\det M = 1$ for $M \in \mathrm{SU}(2)$, the transformation

$$\boldsymbol{X} \longmapsto M \boldsymbol{X} M^\dagger$$ (7.34)

preserves the determinant of $\boldsymbol{X}$, which is now

$$\det \boldsymbol{X} = 0 = \frac{1}{4} - (x^2 + y^2 + z^2). \tag{7.35}$$

But the transformation (7.34) also preserves the trace of $\boldsymbol{X}$, since

$$\operatorname{tr} \boldsymbol{X} = \operatorname{tr}(vv^{\dagger}) = v^{\dagger}v = |v|^2 \tag{7.36}$$

and $M \in \mathrm{SU}(2)$. Such transformations must therefore preserve $x^2 + y^2 + z^2$. Thus, elements of SU(2), acting on $\mathbb{R}^3$ via (7.34), induce transformations in SO(3). Since $-M$ induces the same SO(3) transformation as $+M$, we cannot quite conclude that $\mathrm{SU}(2) \cong \mathrm{SO}(3)$. We say instead that $SU(2)$ is the *double cover* of SO(3). Since the double cover of SO(3) is called Spin(3), we can write this relationship as

$$\mathrm{SU}(2) \cong \mathrm{Spin}(3). \tag{7.37}$$

Under this correspondence, $R_x \in \mathrm{SU}(2)$ does indeed correspond to a counterclockwise rotation by α in the xy plane, and similarly for R_y and R_z, which is why we used the same names as we did for SO(3), and why we introduced the factor of two in (7.22).

7.4 The Geometry of SU(3)

The unitary group in three complex dimensions is defined by

$$\mathrm{SU}(3) = \{M \in \mathbb{C}^{3 \times 3} : M^{\dagger}M = I, \det M = 1\}. \tag{7.38}$$

A (complex) 3×3 matrix has nine complex components, the constraint $M^{\dagger}M = I$ imposes nine real conditions, and the determinant restriction adds just one more (since the other conditions already imply that $|\det M| = 1$). Thus, we expect there to be eight independent "rotations" in SU(3). A set of generators is given by

$$L_1 = L_1(\alpha) = \begin{pmatrix} \cos\left(\frac{\alpha}{2}\right) & -i\sin\left(\frac{\alpha}{2}\right) & 0 \\ -i\sin\left(\frac{\alpha}{2}\right) & \cos\left(\frac{\alpha}{2}\right) & 0 \\ 0 & 0 & 1 \end{pmatrix}, \qquad (7.39)$$

$$L_2 = L_2(\alpha) = \begin{pmatrix} \cos\left(\frac{\alpha}{2}\right) & -\sin\left(\frac{\alpha}{2}\right) & 0 \\ \sin\left(\frac{\alpha}{2}\right) & \cos\left(\frac{\alpha}{2}\right) & 0 \\ 0 & 0 & 1 \end{pmatrix}, \qquad (7.40)$$

$$L_3 = L_3(\alpha) = \begin{pmatrix} e^{-i\alpha/2} & 0 & 0 \\ 0 & e^{i\alpha/2} & 0 \\ 0 & 0 & 1 \end{pmatrix}, \qquad (7.41)$$

$$L_4 = L_4(\alpha) = \begin{pmatrix} \cos\left(\frac{\alpha}{2}\right) & 0 & -i\sin\left(\frac{\alpha}{2}\right) \\ 0 & 1 & 0 \\ -i\sin\left(\frac{\alpha}{2}\right) & 0 & \cos\left(\frac{\alpha}{2}\right) \end{pmatrix}, \qquad (7.42)$$

$$L_5 = L_5(\alpha) = \begin{pmatrix} \cos\left(\frac{\alpha}{2}\right) & 0 & \sin\left(\frac{\alpha}{2}\right) \\ 0 & 1 & 0 \\ -\sin\left(\frac{\alpha}{2}\right) & 0 & \cos\left(\frac{\alpha}{2}\right) \end{pmatrix}, \qquad (7.43)$$

$$L_6 = L_6(\alpha) = \begin{pmatrix} 1 & 0 & 0 \\ 0 & \cos\left(\frac{\alpha}{2}\right) & -i\sin\left(\frac{\alpha}{2}\right) \\ 0 & -i\sin\left(\frac{\alpha}{2}\right) & \cos\left(\frac{\alpha}{2}\right) \end{pmatrix}, \qquad (7.44)$$

$$L_7 = L_7(\alpha) = \begin{pmatrix} 1 & 0 & 0 \\ 0 & \cos\left(\frac{\alpha}{2}\right) & -\sin\left(\frac{\alpha}{2}\right) \\ 0 & \sin\left(\frac{\alpha}{2}\right) & \cos\left(\frac{\alpha}{2}\right) \end{pmatrix}, \qquad (7.45)$$

$$L_8 = L_8(\alpha) = \begin{pmatrix} e^{-i\alpha/2\sqrt{3}} & 0 & 0 \\ 0 & e^{i\alpha/2\sqrt{3}} & 0 \\ 0 & 0 & e^{-i\alpha/\sqrt{3}} \end{pmatrix}, \qquad (7.46)$$

where the factors of 2 and $\sqrt{3}$ are again conventional. These matrices are closely related to the *Gell-Mann matrices*

$$\lambda_m = 2i\frac{\partial L_m}{\partial \alpha} \qquad (7.47)$$

which are explicitly given by[1]

$$\lambda_1 = \begin{pmatrix} 0 & 1 & 0 \\ 1 & 0 & 0 \\ 0 & 0 & 0 \end{pmatrix}, \tag{7.48}$$

$$\lambda_2 = \begin{pmatrix} 0 & -i & 0 \\ i & 0 & 0 \\ 0 & 0 & 0 \end{pmatrix}, \tag{7.49}$$

$$\lambda_3 = \begin{pmatrix} 1 & 0 & 0 \\ 0 & -1 & 0 \\ 0 & 0 & 0 \end{pmatrix}, \tag{7.50}$$

$$\lambda_4 = \begin{pmatrix} 0 & 0 & 1 \\ 0 & 0 & 0 \\ 1 & 0 & 0 \end{pmatrix}, \tag{7.51}$$

$$\lambda_5 = \begin{pmatrix} 0 & 0 & i \\ 0 & 0 & 0 \\ -i & 0 & 0 \end{pmatrix}, \tag{7.52}$$

$$\lambda_6 = \begin{pmatrix} 0 & 0 & 0 \\ 0 & 0 & 1 \\ 0 & 1 & 0 \end{pmatrix}, \tag{7.53}$$

$$\lambda_7 = \begin{pmatrix} 0 & 0 & 0 \\ 0 & 0 & -i \\ 0 & i & 0 \end{pmatrix}, \tag{7.54}$$

$$\lambda_8 = \frac{1}{\sqrt{3}} \begin{pmatrix} 1 & 0 & 0 \\ 0 & 1 & 0 \\ 0 & 0 & -2 \end{pmatrix}. \tag{7.55}$$

The Gell-Mann matrices are again *Hermitian* ($\sigma_m^\dagger = \sigma$) and tracefree ($\text{tr}\,\sigma_m = 0$), but do not share the other properties we listed for the Pauli matrices in the previous section.

The group SU(3) is the smallest of the unitary groups to be unrelated to the orthogonal groups; it's something new.

---

[1]Our definition of λ_5 differs by an overall minus sign from the standard definition, in order to correct a minor but annoying lack of cyclic symmetry in the original definition.

7.5   The Geometry of SU(2, 2)

Just as with the orthogonal groups, we can have unitary groups that preserve an inner product with arbitrary signature. As before, we let g be the diagonal matrix with signature (p, q), and define

$$\mathrm{SU}(p, q) = \{M \in \mathbb{C}^{(p+q) \times (p+q)} : M^\dagger g M = g, \det M = 1\}. \qquad (7.56)$$

If $q = 0$ (or $p = 0$), we recover the ordinary unitary groups. As a nontrivial example, we consider

$$\mathrm{SU}(2, 2) = \{M \in \mathbb{C}^{4 \times 4} : M^\dagger g M = g, \det M = 1\} \qquad (7.57)$$

where g has signature $(2, 2)$.

It is easy to see that $\mathrm{SO}(p, q)$ has precisely pq boosts out of $\frac{1}{2}(n - 1)n$ total elements, where $p + q = n$. A little more work shows that $\mathrm{SU}(p, q)$ has $2pq$ boosts out of $n^2 - 1$ total elements, where again $p + q = n$. Are there cases where these numbers agree (not necessarily for the same values of p, q, and n)?

Two possibilities are

$$\frac{1}{2}(3 - 1)2 = 3 = 2^2 - 1, \qquad (7.58)$$

$$\frac{1}{2}(6 - 1)6 = 15 = 4^2 - 1. \qquad (7.59)$$

With no boosts, that is, with $q = 0$ in both cases, this leads to the identifications

$$\mathrm{SU}(2) \cong \mathrm{Spin}(3), \qquad (7.60)$$

$$\mathrm{SU}(4) \cong \mathrm{Spin}(6) \qquad (7.61)$$

where we must again replace the orthogonal groups by their double covers, the spin groups. We have already seen (7.60), but (7.61) is new. Taking other signatures into account, we have the two further identifications

$$\mathrm{SU}(1, 1) \cong \mathrm{Spin}(2, 1), \qquad (7.62)$$

$$\mathrm{SU}(2, 2) \cong \mathrm{Spin}(4, 2) \qquad (7.63)$$

so that, locally at least, we can identify $\mathrm{SU}(2, 2)$ with $\mathrm{SO}(4, 2)$, which we have already studied.

Knowing now that we can identify elements of $\mathrm{SU}(2, 2)$ with those of $\mathrm{SO}(4, 2)$, we write down generators of $\mathrm{SU}(2, 2)$, using the appropriate names from the generators of $\mathrm{SO}(4, 2)$.

We know that $SO(4,2)$ acts on vectors with six components, such as

$$v = \begin{pmatrix} t \\ x \\ y \\ z \\ p \\ q \end{pmatrix}.$$ (7.64)

Can we find a 6-dimensional representation of $SU(2,2)$? Consider the matrix

$$P = \begin{pmatrix} 0 & t+iq & z+ip & x+iy \\ -t-iq & 0 & -x+iy & z-ip \\ -z-ip & x-iy & 0 & t-iq \\ -x-iy & -z+ip & -t+iq & 0 \end{pmatrix}$$ (7.65)

whose determinant is

$$\det P = (-t^2 + x^2 + y^2 + z^2 + p^2 - q^2)^2$$ (7.66)

which is precisely the (square of the) $SO(4,2)$ (squared) norm of v. Since

$$\det(MPM^T) = \det P$$ (7.67)

so long as $\det M = \pm 1$, elements $M \in SU(2,2)$ will preserve the determinant of P, and are therefore in $O(4,2)$; whether they are in $SO(4,2)$ depends on whether they preserve orientation. Furthermore, P is the most general 4×4 antisymmetric complex matrix, and MPM^T is automatically antisymmetric if P is. Thus, all we need to do to show that $SU(2,2) \subset SO(4,2)$ is to exhibit 15 generators of $SU(2,2)$, and check that they preserve orientations; this also suffices to demonstrate the (local) equivalence between $SU(2,2)$ and $SO(4,2)$, since each has 15 generators.

One possible choice of such generators is given by

$$R_{xy} = \begin{pmatrix} e^{-i\alpha} & 0 & 0 & 0 \\ 0 & e^{i\alpha} & 0 & 0 \\ 0 & 0 & e^{i\alpha} & 0 \\ 0 & 0 & 0 & e^{-i\alpha} \end{pmatrix}, \tag{7.68}$$

$$R_{yz} = \begin{pmatrix} \cos\alpha & i\sin\alpha & 0 & 0 \\ i\sin\alpha & \cos\alpha & 0 & 0 \\ 0 & 0 & \cos\alpha & i\sin\alpha \\ 0 & 0 & i\sin\alpha & \cos\alpha \end{pmatrix}, \tag{7.69}$$

$$R_{zx} = \begin{pmatrix} \cos\alpha & \sin\alpha & 0 & 0 \\ -\sin\alpha & \cos\alpha & 0 & 0 \\ 0 & 0 & \cos\alpha & -\sin\alpha \\ 0 & 0 & \sin\alpha & \cos\alpha \end{pmatrix}, \tag{7.70}$$

$$R_{px} = \begin{pmatrix} \cos\alpha & -i\sin\alpha & 0 & 0 \\ -i\sin\alpha & \cos\alpha & 0 & 0 \\ 0 & 0 & \cos\alpha & i\sin\alpha \\ 0 & 0 & i\sin\alpha & \cos\alpha \end{pmatrix}, \tag{7.71}$$

$$R_{py} = \begin{pmatrix} \cos\alpha & \sin\alpha & 0 & 0 \\ -\sin\alpha & \cos\alpha & 0 & 0 \\ 0 & 0 & \cos\alpha & \sin\alpha \\ 0 & 0 & -\sin\alpha & \cos\alpha \end{pmatrix}, \tag{7.72}$$

$$R_{pz} = \begin{pmatrix} e^{i\alpha} & 0 & 0 & 0 \\ 0 & e^{-i\alpha} & 0 & 0 \\ 0 & 0 & e^{i\alpha} & 0 \\ 0 & 0 & 0 & e^{-i\alpha} \end{pmatrix}, \tag{7.73}$$

$$R_{tx} = \begin{pmatrix} \cosh\alpha & 0 & -\sinh\alpha & 0 \\ 0 & \cosh\alpha & 0 & -\sinh\alpha \\ -\sinh\alpha & 0 & \cosh\alpha & 0 \\ 0 & -\sinh\alpha & 0 & \cosh\alpha \end{pmatrix}, \tag{7.74}$$

$$R_{ty} = \begin{pmatrix} \cosh\alpha & 0 & i\sinh\alpha & 0 \\ 0 & \cosh\alpha & 0 & -i\sinh\alpha \\ -i\sinh\alpha & 0 & \cosh\alpha & 0 \\ 0 & i\sinh\alpha & 0 & \cosh\alpha \end{pmatrix}, \tag{7.75}$$

$$R_{tz} = \begin{pmatrix} \cosh\alpha & 0 & 0 & \sinh\alpha \\ 0 & \cosh\alpha & -\sinh\alpha & 0 \\ 0 & -\sinh\alpha & \cosh\alpha & 0 \\ \sinh\alpha & 0 & 0 & \cosh\alpha \end{pmatrix}, \tag{7.76}$$

$$
R_{tp} = \begin{pmatrix} \cosh\alpha & 0 & 0 & -i\sinh\alpha \\ 0 & \cosh\alpha & -i\sinh\alpha & 0 \\ 0 & i\sinh\alpha & \cosh\alpha & 0 \\ i\sinh\alpha & 0 & 0 & \cosh\alpha \end{pmatrix}, \qquad (7.77)
$$

$$
R_{qx} = \begin{pmatrix} \cosh\alpha & 0 & -i\sinh\alpha & 0 \\ 0 & \cosh\alpha & 0 & -i\sinh\alpha \\ i\sinh\alpha & 0 & \cosh\alpha & 0 \\ 0 & i\sinh\alpha & 0 & \cosh\alpha \end{pmatrix}, \qquad (7.78)
$$

$$
R_{qy} = \begin{pmatrix} \cosh\alpha & 0 & -\sinh\alpha & 0 \\ 0 & \cosh\alpha & 0 & \sinh\alpha \\ -\sinh\alpha & 0 & \cosh\alpha & 0 \\ 0 & \sinh\alpha & 0 & \cosh\alpha \end{pmatrix}, \qquad (7.79)
$$

$$
R_{qz} = \begin{pmatrix} \cosh\alpha & 0 & 0 & i\sinh\alpha \\ 0 & \cosh\alpha & -i\sinh\alpha & 0 \\ 0 & i\sinh\alpha & \cosh\alpha & 0 \\ -i\sinh\alpha & 0 & 0 & \cosh\alpha \end{pmatrix}, \qquad (7.80)
$$

$$
R_{qp} = \begin{pmatrix} \cosh\alpha & 0 & 0 & \sinh\alpha \\ 0 & \cosh\alpha & \sinh\alpha & 0 \\ 0 & \sinh\alpha & \cosh\alpha & 0 \\ \sinh\alpha & 0 & 0 & \cosh\alpha \end{pmatrix}, \qquad (7.81)
$$

$$
R_{tq} = \begin{pmatrix} e^{-i\alpha} & 0 & 0 & 0 \\ 0 & e^{-i\alpha} & 0 & 0 \\ 0 & 0 & e^{i\alpha} & 0 \\ 0 & 0 & 0 & e^{i\alpha} \end{pmatrix}, \qquad (7.82)
$$

and we leave checking the orientation as an exercise. When acting via (7.67), each transformation R_{ab} corresponds to a rotation or boost in the ab plane through an angle (possibly hyperbolic) of 2α.

Chapter 8

Some Symplectic Groups

8.1 Symplectic Transformations

Orthogonal and unitary transformations preserve symmetric inner products; symplectic transformations preserve an antisymmetric product. Let Ω be the $2m \times 2m$ matrix with block structure

$$\Omega = \begin{pmatrix} 0 & I_m \\ -I_m & 0 \end{pmatrix} \tag{8.1}$$

where I_m denotes the $m \times m$ identity matrix. Then the real symmetric groups $\mathrm{Sp}(2m, \mathbb{R})$ are defined by[1]

$$\mathrm{Sp}(2m, \mathbb{R}) = \{M \in \mathbb{R}^{2m \times 2m} : M \Omega M^T = \Omega\}. \tag{8.2}$$

Although not obvious at the group level,[2]

$$M \Omega M^T = \Omega \iff M^T \Omega M = \Omega \tag{8.3}$$

so an equivalent definition is

$$\mathrm{Sp}(2m, \mathbb{R}) = \{M \in \mathbb{R}^{2m \times 2m} : M^T \Omega M = \Omega\}. \tag{8.4}$$

---

[1]There are many different, overlapping conventions for the names of the symplectic groups. The group $\mathrm{Sp}(2m, \mathbb{R})$ is also written as $\mathrm{Sp}(m, \mathbb{R})$. Since $i\Omega$ is a Hermitian inner product of signature (m, m), $\mathrm{Sp}(2m, \mathbb{R})$ is a real subgroup of $\mathrm{SU}(m, m)$, and is also written as $\mathrm{Sp}(m, m)$.

[2]This result is easy to see at the Lie algebra level, that is, by treating M as a 1-parameter family of transformations connected to the identity, and differentiating with respect to the parameter. Denoting the derivative of M at the identity element by A, we have $A\Omega + \Omega A^T = 0$ if $M \in \mathrm{Sp}(2m, \mathbb{R})$. Multiplying on both sides by Ω, and using the fact that $\Omega^2 = -I$ results in $\Omega A + A^T \Omega = 0$ and we have successfully moved the transposed matrix from one side to the other.

Other real forms of the symplectic groups can be obtained by first complexifying $\mathrm{Sp}(2m, \mathbb{R})$, that is, by considering[3]

$$\mathrm{Sp}(2m, \mathbb{R}) \otimes \mathbb{C} = \{M \in \mathbb{C}^{2m \times 2m} : M^T \Omega M = \Omega\}. \tag{8.5}$$

Of particular interest is the compact real form, obtained as the intersection of $\mathrm{Sp}(2m, \mathbb{R}) \otimes \mathbb{C}$ with $\mathrm{SU}(2m)$, that is, the groups

$$\mathrm{Sp}(m) = \{M \in \mathbb{C}^{2m \times 2m} : M^T \Omega M = \Omega, M^\dagger M = I\}. \tag{8.6}$$

The dimension of the symplectic groups is given by

$$|\mathrm{Sp}(m)| = m(2m + 1). \tag{8.7}$$

Real forms with different signatures can be obtained by intersecting with $\mathrm{SU}(p, q)$, where $p + q = 2m$, rather than with $\mathrm{SU}(2m)$.

8.2   The Geometry of Sp(4, ℝ)

From Section 8.1, we have

$$\mathrm{Sp}(4, \mathbb{R}) = \{M \in \mathbb{R}^{4 \times 4} : M \Omega M^T = \Omega\} \tag{8.8}$$

where

$$\Omega = \begin{pmatrix} 0 & 0 & 1 & 0 \\ 0 & 0 & 0 & 1 \\ -1 & 0 & 0 & 0 \\ 0 & -1 & 0 & 0 \end{pmatrix}. \tag{8.9}$$

Consider now the antisymmetric matrix

$$P = \begin{pmatrix} 0 & p+q & -x & t+z \\ -p-q & 0 & z-t & x \\ x & t-z & 0 & p-q \\ -t-z & -x & q-p & 0 \end{pmatrix} \tag{8.10}$$

and note that

$$P^T \Omega P = P \Omega P^T = |P|^2 \Omega \tag{8.11}$$

where

$$|P|^2 = -t^2 + x^2 + z^2 + p^2 - q^2 \tag{8.12}$$

---

[3] Again, there are several conventions for the name of this group, and for the names of its subgroups. However, we avoid the name $\mathrm{Sp}(2m, \mathbb{C})$, which we reserve for a different group.

(and is also the square root of $\det P$). We claim that $M \in \mathrm{Sp}(4, \mathbb{R})$ acts on P via

$$P \longmapsto M^T P M \qquad (8.13)$$

that is, we claim that $M^T P M$ is a matrix of the same form as P. What form is that? P is clearly antisymmetric, and the transformation (8.13) clearly takes antisymmetric matrices to antisymmetric matrices, since

$$(M^T P M)^T = M^T P^T M = -M^T P M. \qquad (8.14)$$

However, antisymmetric matrices in four (real) dimensions contain six independent components, and we only have five. Which degree of freedom are we missing? That's easy: P does not contain a multiple of Ω itself. But that's fine, since (8.13) takes Ω to itself, so if there's no Ω "component" to start with, then there won't be one afterward.

So what does the transformation (8.13) do? We have

$$\begin{aligned}
(M^T P M)\Omega(M^T P M)^T &= M^T P (M\Omega M^T) P^T M \\
&= M^T (P\Omega P^T) M = M^T |P|^2 \Omega M \\
&= |P|^2 \Omega
\end{aligned} \qquad (8.15)$$

since $|P|^2$ is a real number, and therefore commutes with M^T. But $|P|^2$ is just the norm on $\mathbb{R}^5$ with signature $(3, 2)$, which is thus preserved by $\mathrm{Sp}(4, \mathbb{R})$. Since both $\mathrm{Sp}(4, \mathbb{R})$ and $\mathrm{SO}(3, 2)$ are 10-dimensional, we have shown that

$$\mathrm{Sp}(4, \mathbb{R}) \cong \mathrm{Spin}(3, 2) \qquad (8.16)$$

where $\mathrm{Spin}(3, 2)$ is of course the double cover of $\mathrm{SO}(3, 2)$.

8.3 The Geometry of Sp(6, ℝ)

It is instructive to compare the description of $\mathrm{Sp}(4, \mathbb{R}) \cong \mathrm{Spin}(3, 2)$ given in Section 8.2 with the description of $\mathrm{SU}(2, 2) \cong \mathrm{Spin}(4, 2)$ given in Section 7.5. It is clear that $\mathrm{Sp}(4, \mathbb{R})$ is (can be identified with) a subgroup of $\mathrm{SU}(2, 2)$; it turns out to be the "real part" of $\mathrm{SU}(2, 2)$. In other words, $\mathrm{SU}(2, 2)$ turns out to deserve the name "$\mathrm{Sp}(4, \mathbb{C})$." A similar construction shows that $\mathrm{Sp}(6, \mathbb{R})$ can be interpreted as the real part of $\mathrm{SU}(3, 3)$, which in turn deserves the name "$\mathrm{Sp}(6, \mathbb{C})$."

As discussed in Section 9.4, this usage requires a choice of what one means by "symplectic", after which the identifications above are straightforward. It should however be emphasized that this correspondence between unitary and symplectic groups fails over $\mathbb{H}$ and $\mathbb{O}$.

Further insight into the structure of $\mathrm{Sp}(6, \mathbb{C})$ can be obtained by reading Section 11.5, but replacing $\mathbb{O}$ everywhere by $\mathbb{C}$.

Chapter 9

Symmetry Groups over Other Division Algebras

9.1 Some Orthogonal Groups over Other Division Algebras

Each of the division algebras corresponds to a $2k$-dimensional vector space, with positive-definite inner product. Multiplication by unit-normed elements preserves the norm, and thus induces either a rotation or a reflection on the vector space. We therefore expect to be able to represent several orthogonal groups in terms of division algebra multiplication. This expectation is correct, but there are several cases, including some surprises due to the lack of commutativity and associativity.

9.1.1 *A Quaternionic Description of* SO(3)

We saw in Section 7.2 how to generate SO(2) through multiplication by a unit-normed complex number. We can try the same thing with the quaternions: Multiplication by a unit-normed quaternion should be a rotation. Since the quaternions are 4-dimensional, we expect to get rotations in four dimensions.

Not so fast! We saw in Section 3.4 that quaternionic multiplication corresponds to rotations in *two* planes, making it difficult to count. Let's try to generate rotations in just one plane.

We start with the *imaginary* quaternions, which correspond to vectors in *three* dimensions. If we want to rotate only these three dimensions, we need transformations that leave the real part of a quaternion alone. Consider the map

$$q \longmapsto pqp^{-1} \tag{9.1}$$

which is called *conjugation* of q by p.[1] If $[p, q] = 0$, that is, if p and q commute, then of course $pqp^{-1} = q$. In particular, conjugation preserves 1, and hence all real numbers.

Conjugation automatically preserves the norm of q, since

$$|pqp^{-1}| = |p|\,|q|\,|p|^{-1} = |q|. \tag{9.2}$$

This argument shows that the norm of p plays no role in the transformation, so we can assume $|p| = 1$. So let $p = e^{i\alpha}$. Then $p^{-1} = \overline{p}$, and conjugation by p becomes

$$q \longmapsto pq\overline{p}. \tag{9.3}$$

What does conjugation by p do to q? Since $\mathbb{H} = \mathbb{C} \oplus \mathbb{C}j$, we can write

$$q = r_1 e^{i\theta} + r_2 e^{i\phi}j. \tag{9.4}$$

Remembering that i and j anticommute, we therefore have

$$pq\overline{p} = e^{i\alpha}\left(r_1 e^{i\theta} + r_2 e^{i\phi}j\right)e^{-i\alpha} = r_1 e^{i\theta} + r_2 e^{i(\phi+2\alpha)}j. \tag{9.5}$$

Thus, conjugation by $e^{i\alpha}$ leaves both the real and i directions alone, but induces a rotation by 2α in the jk plane. Similarly, conjugation by $e^{j\alpha}$ and $e^{k\alpha}$ correspond to rotations by 2α in the ki and ij planes, respectively. Since we can generate any rotation by combining rotations in the coordinate planes, we see that conjugation by unit-normed quaternions generates $SO(3)$, the rotations in $\mathrm{Im}\,\mathbb{H}$.[2] In three dimensions, we can actually do better: *any* rotation in three dimensions is in fact a rotation about a single axis, so we can dispense with the notion of "generators" in this case. In other words,

$$SO(3) = \{p \in \mathbb{H} : |p| = 1\} \tag{9.6}$$

or, in the sense of transformations,

$$SO(3) = \{q \longmapsto pq\overline{p} : p, q \in \mathbb{H}, |p| = 1\}. \tag{9.7}$$

A special case occurs when $p = i$, corresponding to a rotation by π in the jk plane, which takes every element of the jk plane and multiplies it by -1. We call such a transformation a *flip*. Any imaginary unit u can be used here, corresponding to a flip in the plane perpendicular to it.

What about one-sided multiplication by unit-normed quaternions? These transformations are in one-to-one correspondence with conjugation,

---

[1]Do not confuse the two different uses of the word "conjugation"!

[2]As noted above, the restriction on the norm can be dropped, since it cancels out of the transformation.

so there aren't enough transformations to generate all of SO(4). In fact, this must be another version of SO(3)—not the same version, since the identity element is not left invariant. Let's take a closer look.

Starting again from (9.4), we have

$$e^{i\alpha} \left(r_1 e^{i\theta} + r_2 e^{i\phi} j \right) = r_1 e^{i(\theta+\alpha)} + r_2 e^{i(\phi+\alpha)} j \qquad (9.8)$$

so that left multiplication by $e^{i\alpha}$ corresponds to a rotation by α in both the $1i$ and jk planes. What about right multiplication? Now we have

$$\left(r_1 e^{i\theta} + r_2 e^{i\phi} j \right) e^{i\alpha} = r_1 e^{i(\theta+\alpha)} + r_2 e^{i(\phi-\alpha)} j \qquad (9.9)$$

corresponding to a rotation by $+\alpha$ in the $1i$ plane, but a rotation by $-\alpha$ in the jk plane.

We conclude that there are *three* different versions of SO(3) here, induced respectively by conjugation, left multiplication, and right multiplication by a unit quaternion.

9.1.2 *A Quaternionic Description of* SO(4)

We return to the question of finding rotations in four dimensions. Conjugation doesn't do the job—as we saw in Section 9.1.1, conjugation induces rotations in three dimensions, and single-sided multiplication isn't sufficient to induce rotations in four dimensions. Therein lies a clue; what about double-sided multiplication?

We compute

$$e^{i\alpha} \left(r_1 e^{i\theta} + r_2 e^{i\phi} j \right) e^{i\alpha} = r_1 e^{i\theta+2\alpha} + r_2 e^{i\phi} j \qquad (9.10)$$

corresponding to a rotation by 2θ in the $1i$ plane. Again, we can repeat this construction with i replaced by j and k to produce rotations in the $1j$ and $1k$ planes, respectively. Combining these rotations with those in the imaginary planes (ij, jk, and ki), we have constructed rotations in each of the six coordinate planes, from which arbitrary rotations can be constructed.

We have shown that every transformation in SO(4) is a combination of transformations that take q either to $pq\bar{p} = pqp^{-1}$ or pqp. However, not every SO(4) transformation is necessarily of this form, so we write

$$\text{SO}(4) = \langle \{q \longmapsto pq\bar{p} : p, q \in \mathbb{H}, |p| = 1\}$$
$$\cup \{q \longmapsto pqp : p, q \in \mathbb{H}, |p| = 1\} \rangle \qquad (9.11)$$

where the angled brackets imply that SO(4) is *generated* by the given transformations.

Thanks to the associativity of the quaternions, we can rewrite these transformations in several ways. First of all, we can combine all of the factors on the left into a single factor, and the same on the right. Thus,

$$\mathrm{SO}(4) = \langle \{ q \longmapsto p_1 q p_2 : p_1, p_2, q \in \mathbb{H}, |p_1| = 1 = |p_2| \}. \tag{9.12}$$

Furthermore, if we first conjugate q by p, then multiply by p on both sides, we could equivalently have multiplied by p^2 on the left; a similar construction yields multiplication on the right. Thus, instead of combining conjugation and two-sided multiplication, we could equivalently have combined left and right multiplication, that is

$$\mathrm{SO}(4) = \langle \{ q \longmapsto pq : p \in \mathbb{H}, |p| = 1 \} \cup \{ q \longmapsto qp : p \in \mathbb{H}, |p| = 1 \} \rangle. \tag{9.13}$$

But we have already seen that single-sided multiplication generates a version of $\mathrm{SO}(3)$. We have therefore established an equivalence between two copies of $\mathrm{SO}(3)$ and $\mathrm{SO}(4)$, which is customarily written as

$$\mathrm{SO}(4) \cong \mathrm{SO}(3) \times \mathrm{SO}(3). \tag{9.14}$$

9.1.3 *An Octonionic Description of* $\mathrm{SO}(7)$

We showed in Section 9.1.1 that conjugation by a unit-normed imaginary quaternion u yields a flip of imaginary quaternions about the u-axis. Flips cannot only be used with quaternions, but also with octonions, since the expression $px\bar{p}$ involves only two directions, and hence lies in a quaternionic subalgebra of $\mathbb{O}$; there are no associativity issues here.

But *any* rotation can be constructed from two flips. For instance, to rotate the xy-plane, first pick any line (through the origin) in that plane. Now rotate by π about that line, that is, take all points in the plane orthogonal to the given line by -1. For example, if the chosen line is the x-axis, then the x-coordinate of a point is unaffected, while its y- and z-coordinates are multiplied by -1. Now pick another line in the xy-plane, at an angle α from the first line, and repeat the process. Points along the z axis are reflected twice, and are thus taken back to where they started. But any point in the xy-plane winds up being rotated by 2α! (This is easiest to see for points along the x-axis.)

It doesn't matter which two lines in the xy-plane we choose, so long as they are separated by α (with the correct orientation). And we have described this procedure as though it were taking place in three dimensions, but in fact it works in any number of dimensions; there can be any number of "z-coordinates", all of which are flipped twice, and return to where they started.

To rotate counterclockwise by an angle 2α in the ij-plane, we therefore begin by conjugating with i, thus rotating about the i-axis. To complete the ij rotation, we need to rotate about the line in the ij-plane which makes an angle α with the i-axis. This is accomplished by conjugating by a unit octonion u pointing along the line, which is easily seen to be

$$u = i \cos\alpha + j \sin\alpha. \tag{9.15}$$

Finally, note that the conjugate of any imaginary octonion is just minus itself. Putting this all together, a rotation by 2α in the ij-plane is given by

$$x \longmapsto (i \cos\alpha + j \sin\alpha)(ixi)(i \cos\alpha + j \sin\alpha) \tag{9.16}$$

for any octonion x. (We have removed two minus signs.)

If $x \in \mathbb{H}$, we can collapse the parentheses in (9.16), obtaining

$$x \longmapsto (-\cos\alpha - k\sin\alpha)x(-\cos\alpha + k\sin\alpha) = e^{k\alpha}xe^{-k\alpha} \tag{9.17}$$

which is just conjugation by a unit quaternion, as in the construction of SO(3) in Section 9.1.1. Over the octonions, however, we cannot simplify (9.16) any further; it takes *two* transformations, not just one, to rotate a single plane. We refer to this process as *nesting*, and describe the transformation (9.16) as a *nested flip*.

We can repeat this construction using any unit-normed imaginary units u, v that are orthogonal to each other, obtaining the rotation in the uv-plane. Since such rotations generate SO(7), we have

$$\mathrm{SO}(7) = \langle \{ x \longmapsto (u\cos\alpha + v\sin\alpha)(uxu)(u\cos\alpha + v\sin\alpha) :$$
$$u, v, x \in \mathbb{O}, u^2 = -1 = v^2, \{u, v\} = 0 \} \rangle. \tag{9.18}$$

9.1.4 *An Octonionic Description of* SO(8)

The transition from SO(7) to SO(8) is much the same as from SO(3) to SO(4): Flips generate SO(7), and two-sided multiplication generates the rotations with the real direction. To see the latter property, consider the transformation

$$x \longmapsto e^{\ell\alpha}xe^{\ell\alpha}. \tag{9.19}$$

We can separate $x \in \mathbb{O}$ into a piece in the complex subalgebra containing ℓ, and a piece orthogonal to this subalgebra. That is, since

$$\mathbb{O} = \mathbb{C} \oplus \mathbb{C}^{\perp} \tag{9.20}$$

we have

$$x = re^{\ell\theta} + x^{\perp} \tag{9.21}$$

where $x^\perp$ is orthogonal to ℓ. Recall that imaginary octonions are orthogonal if they anticommute, that is

$$x \perp y \iff \{x, y\} = xy + yx = 0 \tag{9.22}$$

so long as $\mathrm{Re}\,(x) = 0 = \mathrm{Re}\,(y)$, and that

$$e^{u\alpha} y = y e^{-u\alpha} \tag{9.23}$$

if $u \perp y$ (and again both u and y are imaginary). Thus,

$$e^{\ell\alpha} x e^{\ell\alpha} = e^{\ell\alpha}(re^{\ell\theta} + x^\perp)e^{\ell\alpha} = re^{\ell\theta+2\alpha} + x^\perp \tag{9.24}$$

where we have used (9.23).

Without further ado, we have

$$\begin{aligned}
\mathrm{SO}(8) = \langle\{x \longmapsto (u\cos\alpha + v\sin\alpha)(uxu)(u\cos\alpha + v\sin\alpha) : \\
u, v, x \in \mathbb{O}, u^2 = -1 = v^2, \{u, v\} = 0\} \\
\cup \{x \longmapsto pxp : p, x \in \mathbb{O}, |p| = 1\}\rangle.
\end{aligned} \tag{9.25}$$

However, *both* of these types of transformations are (generated by) symmetric multiplication. Furthermore, unlike with the quaternions, single-sided multiplication with the octonions actually generates all of $\mathrm{SO}(8)$, not merely a subset of it. This is an important property of $\mathrm{SO}(8)$, known as *triality*, which says that each of the following representations is equivalent:[3]

$$\mathrm{SO}(8) = \langle\{x \longmapsto pxp : p, x \in \mathbb{O}, |p| = 1\}\rangle, \tag{9.26}$$
$$\mathrm{SO}(8) = \langle\{x \longmapsto px : p, x \in \mathbb{O}, |p| = 1\}\rangle, \tag{9.27}$$
$$\mathrm{SO}(8) = \langle\{x \longmapsto xp : p, x \in \mathbb{O}, |p| = 1\}\rangle. \tag{9.28}$$

This difference between $\mathrm{SO}(8)$ and $\mathrm{SO}(4)$ is due to nesting. Single-sided multiplication *generates* $\mathrm{SO}(8)$; unlike over the quaternions, iterated multiplications do not collapse to a single operation over the octonions.

9.2   Some Unitary Groups over Other Division Algebras

In Section 7.3, the complex group $\mathrm{SU}(2) = \mathrm{SU}(2, \mathbb{C})$ was represented in terms of 2×2 matrices R_x, R_y, R_z, acting on complex matrices of the

---

[3]More formally, (9.26) yields the *vector* representation of $\mathrm{SO}(8)$, (9.27) yields the *spinor* representation of $\mathrm{SO}(8)$, and (9.28) yields the *dual spinor* representation of $\mathrm{SO}(8)$, in all cases acting on the 8-dimensional space $\mathbb{O}$. *Triality* is the implicit map between these three representations given by relating the transformations determined by the same element p.

form[4]

$$X = \begin{pmatrix} 1 + z & x - iy \\ x + iy & 1 - z \end{pmatrix} \tag{9.29}$$

thus demonstrating the isomorphism $\mathrm{SU}(2) \cong \mathrm{Spin}(3)$, the double cover of $\mathrm{SO}(3)$. Recall that (setting $\beta = \frac{\alpha}{2}$ for convenience)

$$R_y = \begin{pmatrix} \cos\beta & -\sin\beta \\ \sin\beta & \cos\beta \end{pmatrix} \tag{9.30}$$

is real, whereas R_x and R_z are complex. But R_y has precisely the form of an element of $\mathrm{SO}(2)$, as discussed in Section 6.2. We therefore identify $\mathrm{SO}(2)$ as a "real unitary" matrix, that is, we write

$$\mathrm{SU}(2, \mathbb{R}) \cong \mathrm{SO}(2). \tag{9.31}$$

Can we go in the other direction? Extending X is easy; just replace the complex number by a division algebra element, so that now

$$X = \begin{pmatrix} 1 + z & \overline{a} \\ a & 1 - z \end{pmatrix} \tag{9.32}$$

with $a \in \mathbb{K}$.

It is important to realize that X is *complex*, in the sense that it involves only one octonionic direction. In particular,

$$\det X = 1 - |a|^2 - z^2. \tag{9.33}$$

Since $\operatorname{tr} X = 2$ and does not involve z or a, any transformation that preserves both the determinant and trace of X will also preserve the norm $|a|^2 + z^2$. We therefore expect

$$\mathrm{SU}(2, \mathbb{H}) \cong \mathrm{Spin}(5), \tag{9.34}$$
$$\mathrm{SU}(2, \mathbb{O}) \cong \mathrm{Spin}(9), \tag{9.35}$$

where the Spin groups are of course the double covers of the corresponding orthogonal groups. But which transformations are these?

Over $\mathbb{H}$, this question is easy to answer. As in Section 7.3, consider transformations of the form

$$X \longmapsto M X M^\dagger \tag{9.36}$$

and build M from R_x, R_y, and R_z, but allow i to be replaced by j or k in R_x and R_y. As is easily checked by direct computation, each of these seven

---

[4]This is really $2X$, with x, y, z rescaled for convenience.

transformations preserves the determinant and trace of $\boldsymbol{X}$. But there are $\binom{5}{2} = 10$ independent generators of Spin(5). Which ones are we missing? We've left out the transformations that mix up i, j, k. But we know from Section 9.1.1 how to implement these transformations using conjugation. Now, however, we must conjugate using multiples of the (2×2) identity matrix $\boldsymbol{I}$, such as the *phase transformation*

$$\boldsymbol{M} = e^{i\alpha} = e^{i\alpha}\boldsymbol{I}. \tag{9.37}$$

Since conjugation leaves real numbers alone, the diagonal of $\boldsymbol{X}$ is not affected. These are our three missing generators, and we have indeed verified (9.34).

The same procedure works over $\mathbb{O}$ as well, but we must be careful about associativity. We now have seven versions of R_x and R_z, which together with R_y yield 15 generators. Which generators are we missing? The 21 SO(7) transformations that mix up (only) the imaginary units. But there are only seven phase transformations of the form (9.37)!

The resolution to this apparent quandary is to recall the discussion in Section 9.1.3, where it was shown that *two* flips must be nested in order to generate rotations in just one plane; the same principle applies to phase transformations. Thus, there are in fact $\binom{7}{2} = 21$ (nested!) phase transformations, just what we need to establish (9.35).

9.3   Some Lorentz Groups over Other Division Algebras

The construction in Section 9.2 can be extended to the corresponding Lorentz groups, as was originally done by Sudbery [4].

We normally think of a vector in Minkowski spacetime in the form

$$\boldsymbol{x} = \begin{pmatrix} t \\ x \\ y \\ z \end{pmatrix}. \tag{9.38}$$

But there is another natural way to package these four degrees of freedom. Consider the complex matrix

$$\boldsymbol{X} = \begin{pmatrix} t + z & x - iy \\ x + iy & t - z \end{pmatrix} = t\,\boldsymbol{I} + x\,\boldsymbol{\sigma}_x + y\,\boldsymbol{\sigma}_y + z\,\boldsymbol{\sigma}_z. \tag{9.39}$$

This matrix is *Hermitian*, that is

$$\boldsymbol{X}^{\dagger} = \boldsymbol{X} \tag{9.40}$$

where the Hermitian conjugate of a matrix, denoted by a dagger, is the matrix obtained by taking both the transpose and the (complex) conjugate of the original matrix. Note that 2×2 complex Hermitian matrices have precisely four (real) independent components; the diagonal elements must be real, and the off-diagonal elements must be (complex) conjugates of each other.

As discussed in Section 6.6, *Lorentz transformations* are transformations which preserve the "squared length" of $\boldsymbol{x}$,

$$|\boldsymbol{x}|^2 = x^2 + y^2 + z^2 - t^2 \tag{9.41}$$

and these transformations form the Lorentz group, $SO(3,1)$.

How can we express such transformations in terms of the matrix $\boldsymbol{X}$? The beauty of this description lies in the fact that the norm of $\boldsymbol{x}$ is just (minus) the determinant of $\boldsymbol{X}$, that is

$$- \det \boldsymbol{X} = |\boldsymbol{x}|^2 = x^2 + y^2 + z^2 - t^2. \tag{9.42}$$

We therefore seek linear transformations which preserve the determinant.

At first sight, this is easy. Since (complex) determinants satisfy

$$\det(\boldsymbol{X}\boldsymbol{Y}) = (\det \boldsymbol{X})(\det \boldsymbol{Y}) \tag{9.43}$$

it would seem that all we need to do is multiply $\boldsymbol{X}$ (on either side) by a matrix with determinant 1. There are two problems with this approach. First of all, there is no simple condition to ensure that the product $\boldsymbol{X}\boldsymbol{Y}$ is Hermitian! Furthermore, this approach will fail over the other division algebras; (9.43) is not true for quaternionic matrices.

We already know how to solve these problems. We consider conjugation of $\boldsymbol{X}$ by a matrix $\boldsymbol{M}$, that is, we consider the transformation

$$\boldsymbol{X} \longmapsto \boldsymbol{M}\boldsymbol{X}\boldsymbol{M}^\dagger. \tag{9.44}$$

Using the property

$$(\boldsymbol{X}\boldsymbol{Y})^\dagger = \boldsymbol{Y}^\dagger \boldsymbol{X}^\dagger \tag{9.45}$$

it is easy to check that $\boldsymbol{M}\boldsymbol{X}\boldsymbol{M}^\dagger$ is Hermitian if $\boldsymbol{X}$ is—with no such restriction on $\boldsymbol{M}$. We therefore seek complex matrices $\boldsymbol{M}$ such that

$$\det(\boldsymbol{M}\boldsymbol{X}\boldsymbol{M}^\dagger) = \det(\boldsymbol{M}) \det(\boldsymbol{X}) \det(\boldsymbol{M}^\dagger) = \det \boldsymbol{X} \tag{9.46}$$

or equivalently

$$\det(\boldsymbol{M}\boldsymbol{M}^\dagger) = \det(\boldsymbol{M}) \det(\boldsymbol{M}^\dagger) = 1.$$

Since

$$\det \boldsymbol{M}^\dagger = \overline{\det \boldsymbol{M}} \tag{9.47}$$

we must have

$$|\det \boldsymbol{M}| = 1. \tag{9.48}$$

However, over $\mathbb{C}$, we can assume without loss of generality that $\det \boldsymbol{M} = 1$, since we can always multiply $\boldsymbol{M}$ by a complex phase without affecting anything else. We have in fact shown that the group of complex matrices with determinant 1, written as $\mathrm{SL}(2, \mathbb{C})$, is (locally) the same as the Lorentz group. In fact,

$$\mathrm{SL}(2, \mathbb{C}) \cong \mathrm{Spin}(3, 1) \tag{9.49}$$

the double cover of $\mathrm{SO}(3, 1)$.

What do these transformations look like? What matrices $\boldsymbol{M}$ correspond to spatial rotations? The spatial rotations can be built from the $\mathrm{SU}(2)$ transformations R_x, R_y, and R_z, as discussed in Section 7.3. Because of the importance of the Lorentz group, we repeat that construction here, using slightly different notation.

First of all, a rotation by 2θ in the xy-plane is given by

$$\boldsymbol{R}_z = \begin{pmatrix} e^{-i\theta} & 0 \\ 0 & e^{i\theta} \end{pmatrix}. \tag{9.50}$$

It is straightforward to check that if

$$\boldsymbol{X}' = \boldsymbol{R}_z \boldsymbol{X} \boldsymbol{R}_z^\dagger = \begin{pmatrix} t' + z' & x' - iy' \\ x' + iy' & t' - z' \end{pmatrix} \tag{9.51}$$

then

$$
\begin{aligned}
t' &= t, \\
x' &= x \cos 2\theta - y \sin 2\theta, \\
y' &= x \sin 2\theta + y \cos 2\theta, \\
z' &= z,
\end{aligned}
\tag{9.52}
$$

which corresponds to a counterclockwise rotation by 2θ in the xy-plane as claimed. Of course, (9.52) can be put in matrix form, as

$$\boldsymbol{x}' = \boldsymbol{\Lambda} \boldsymbol{x} \tag{9.53}$$

from which the traditional representation of this Lorentz transformation as a 4×4 matrix $\boldsymbol{\Lambda}$ can easily be determined. Similarly, rotations in the yz- and zx-planes are given, respectively, by

$$\boldsymbol{R}_x = \begin{pmatrix} \cos\theta & -i\sin\theta \\ -i\sin\theta & \cos\theta \end{pmatrix}, \qquad \boldsymbol{R}_y = \begin{pmatrix} \cos\theta & -\sin\theta \\ \sin\theta & \cos\theta \end{pmatrix}. \tag{9.54}$$

Any rotation can be built up out of these generators.

What about the other Lorentz transformations, namely the ones which "rotate" the time axis? These transformations, called *boosts*, are at the heart of special relativity; geometrically boosts are simply hyperbolic rotations. A boost in the zt-plane takes the form

$$\boldsymbol{B}_z = \begin{pmatrix} e^\beta & 0 \\ 0 & e^{-\beta} \end{pmatrix}. \tag{9.55}$$

It is straightforward to check that if

$$\boldsymbol{X}' = \boldsymbol{B}_z \boldsymbol{X} \boldsymbol{B}_z^\dagger \tag{9.56}$$

then

$$
\begin{aligned}
t' &= t \cosh 2\beta + z \sinh 2\beta, \\
x' &= x, \\
y' &= y, \\
z' &= t \sinh 2\beta + z \cosh 2\beta,
\end{aligned}
\tag{9.57}
$$

corresponding to a boost in the zt-plane by 2β, that is, to a relative speed of $c \tanh 2\beta$. Similarly, boosts in the xt- and yt-planes are given, respectively, by

$$\boldsymbol{B}_x = \begin{pmatrix} \cosh \beta & \sinh \beta \\ \sinh \beta & \cosh \beta \end{pmatrix}, \qquad \boldsymbol{B}_y = \begin{pmatrix} \cosh \beta & -i \sinh \beta \\ i \sinh \beta & \cosh \beta \end{pmatrix}. \tag{9.58}$$

What happens over the other division algebras? As in Section 9.2, simply replace the complex number $x + iy$ in $\boldsymbol{X}$ by a division algebra element a, so that

$$\boldsymbol{X} = \begin{pmatrix} t+z & \overline{a} \\ a & t-z \end{pmatrix}. \tag{9.59}$$

Since a has 1, 2, 4, or 8 components depending on whether a is in $\mathbb{R}$, $\mathbb{C}$, $\mathbb{H}$, or $\mathbb{O}$, respectively, $\boldsymbol{X}$ corresponds to a vector in a spacetime with 3, 4, 6, or 10 dimensions. As before, the determinant gives the Lorentzian norm

$$-\det \boldsymbol{X} = |a|^2 + z^2 - t^2. \tag{9.60}$$

There is no problem defining the determinant here, since even in the octonionic case the components of $\boldsymbol{X}$ lie in a complex subalgebra of $\mathbb{O}$.

We are therefore led to seek transformations of the form (9.44) which preserve the determinant. Even over the quaternions, however, we immediately have a problem: As already noted, (9.43) is not true for quaternionic matrices! Furthermore, it is not at all obvious how to define the determinant in the first place for non-Hermitian matrices.

Fortunately, there is another identity which comes to the rescue here.

$$\det(\boldsymbol{M}\boldsymbol{X}\boldsymbol{M}^\dagger) = \det(\boldsymbol{M}^\dagger\boldsymbol{M})\det\boldsymbol{X} \tag{9.61}$$

so we need to look for quaternionic matrices $\boldsymbol{M}$ satisfying

$$\det(\boldsymbol{M}^\dagger\boldsymbol{M}) = 1. \tag{9.62}$$

The order doesn't matter here, since

$$\det(\boldsymbol{M}\boldsymbol{M}^\dagger) = \det(\boldsymbol{M}^\dagger\boldsymbol{M}) \tag{9.63}$$

holds over $\mathbb{H}$.

Over the octonions, the situation is even worse: Because of the lack of associativity, (9.44) is not well-defined—nor is it clear that the right-hand side is Hermitian! We resolve this difficulty by restricting to those matrices $\boldsymbol{M}$ for which $\boldsymbol{M}\boldsymbol{X}\boldsymbol{M}^\dagger$ is well-defined for all Hermitian $\boldsymbol{X}$. It turns out to be sufficient to assume that the components of $\boldsymbol{M}$ lie in a complex subalgebra of $\mathbb{O}$.[5] For such matrices, the transformation (9.44) involves only two independent directions, and is therefore quaternionic. In particular, (9.61) will hold.

It appears to be straightforward to generalize the rotations and boosts given above. $\boldsymbol{R}_y$ still rotates the real direction in a with z; $\boldsymbol{B}_z$ still yields a boost in the z-direction; $\boldsymbol{B}_x$ still yields a boost in the x-direction. The remaining transformations are nearly as easy. $\boldsymbol{B}_y$ yields a boost in the i-direction, with obvious generalizations to the other spatial directions obtained by replacing i by j, k, etc. Similarly, $\boldsymbol{R}_x$ and $\boldsymbol{R}_z$ yield rotations in the plane defined by either the real part of a or z, respectively, and the i-direction—again generalizing to j, k, etc.

Counting up what we've got, we see that we have boosts in all directions, as well as all rotations involving either z or the real part of a. That's more than enough to generate all the Lorentz transformations *except* those involving two imaginary directions in a. But we know how to do these!

In Section 9.1.1, we obtained a rotation in the jk-plane for single quaternions by conjugating with $e^{i\theta}$. That works here as well! Conjugating $\boldsymbol{X}$ by the matrix

$$\boldsymbol{R}_i = \begin{pmatrix} e^{i\theta} & 0 \\ 0 & e^{i\theta} \end{pmatrix} = e^{i\theta}\boldsymbol{I} \tag{9.64}$$

---

[5]The only other possibility is for the columns of the imaginary part of $\boldsymbol{M}$ to be real multiples of each other [5].

won't touch the diagonal of $\boldsymbol{X}$, but will precisely rotate a (and $\bar{a}$!) by 2θ in the jk-plane. Similarly, replacing i with j and k yields rotations in the other imaginary planes. The phase freedom in the complex case has become an essential ingredient in performing these "internal" rotations over $\mathbb{H}$!

Over the octonions, we must be a bit more careful. As noted in Sections 9.1.3 and 9.2, conjugation by $e^{i\theta}$, and hence by $\boldsymbol{R}_i$, rotates three planes, not just one. But again, we know how to solve this problem: Use flips! For instance,

$$\boldsymbol{X} \longmapsto (i\cos\theta + j\sin\theta)i\boldsymbol{X}i(i\cos\theta + j\sin\theta) \tag{9.65}$$

is a rotation by an angle 2θ in the ij-plane. Yet again, we see that the lack of associativity has come to the rescue; the ability to *nest* transformations is crucial to this construction.

We have therefore obtained an explicit form for the generators of Lorentz transformations in 3, 4, 6, and 10 dimensions. But we have actually shown more, namely that

$$\mathrm{SL}(2,\mathbb{R}) \cong \mathrm{Spin}(2,1), \tag{9.66}$$

$$\mathrm{SL}(2,\mathbb{C}) \cong \mathrm{Spin}(3,1), \tag{9.67}$$

$$\mathrm{SL}(2,\mathbb{H}) \cong \mathrm{Spin}(5,1), \tag{9.68}$$

$$\mathrm{SL}(2,\mathbb{O}) \cong \mathrm{Spin}(9,1), \tag{9.69}$$

that is, $\mathrm{SL}(2,\mathbb{K})$ is the double cover of the Lorentz group $\mathrm{SO}(k+1,1)$, where $k = |\mathbb{K}| = 1,2,4,8$. However, the groups $\mathrm{SL}(2,\mathbb{H})$ and $\mathrm{SL}(2,\mathbb{O})$ require some explanation. The notation "SL" normally means those matrices with determinant 1, but the determinant of a quaternionic matrix is not well-defined. The generalization? To require precisely (9.62)! Over the octonions, all we must do is to also add the restriction that (9.44) be well-defined.[6]

In what sense do the Lorentz groups as defined here consist of "all matrices of determinant 1"? In the complex case, multiplication of M by an arbitrary phase $e^{i\theta}$ does not change the action (9.44), so we can safely restrict to matrices with $\det M = 1$. The quaternionic case is more subtle:

---

[6]Wait a minute, how can $\mathrm{SL}(2,\mathbb{O})$ be a group when $\mathbb{O}$ isn't associative? Simple; the multiplication in $\mathrm{SL}(2,\mathbb{O})$ is not matrix multiplication, but composition, that is

$$(\boldsymbol{M}_1 \bullet \boldsymbol{M}_2)[\boldsymbol{X}] = \boldsymbol{M}_1[\boldsymbol{M}_2[\boldsymbol{X}]] = \boldsymbol{M}_1(\boldsymbol{M}_2\boldsymbol{X}\boldsymbol{M}_2^\dagger)\boldsymbol{M}_1^\dagger$$

where we have used square brackets to denote the action of an element of the group on a vector, and a bullet to indicate the group operation, which is associative. Similar comments apply to other groups constructed over $\mathbb{O}$, such as $\mathrm{SU}(2,\mathbb{O})$.

Multiplication by $e^{i\theta}$ now corresponds to a rotation in the jk-plane, and therefore must be included as a Lorentz transformation even though its determinant is not real. However, we can rewrite such phase transformations as a product of two flips, each with determinant -1, and this construction carries over unchanged to the octonionic case. Furthermore, it is straightforward to rewrite the remaining Lorentz generators, which are already matrices with determinant $+1$, as the product of two matrices each with determinant -1.

Thus, the Lorentz groups could be defined for each of the division algebras as being generated by those transformations consisting of *two* complex matrices of determinant -1, which suitably generalizes the more traditional definition in terms of matrices of determinant $+1$. It is only in this nested sense that $SL(2, \mathbb{O})$ consists of "all matrices of determinant $+1$."

9.4   Some Symplectic Groups over Other Division Algebras

In Section 8.1 we showed that $Sp(2) \cong Spin(5)$, the double cover of $SO(5)$, and in Section 9.2 we showed that $SU(2, \mathbb{H}) \cong Spin(5)$. Meanwhile, in Section 8.2 we showed that $Sp(4, \mathbb{R}) \cong Spin(3, 2)$, the double cover of $SO(3, 2)$. Is there a relationship between symplectic groups and the quaternions?

We first ask under what conditions complex matrices can be reinterpreted as $m \times m$ *quaternionic* matrices. It is easier to go the other way: How do we turn a quaternionic matrix into a complex matrix?

Begin with the simplest case: How do we turn a quaternion into a 2×2 complex matrix? That's easy: Use ($\pm i$ times the) Pauli matrices to represent the quaternionic units. Thus, the matrix representation of the quaternion $a + bi + cj + dk$ is given by

$$aI + b(i\sigma_z) - c(i\sigma_y) + d(i\sigma_x) = \begin{pmatrix} a + bi & -c + di \\ c + di & a - bi \end{pmatrix}. \tag{9.70}$$

This decomposition is just the Cayley–Dickson process in reverse!

What properties do such matrices have? It's not hard to see that a complex matrix M has the form given in (9.70) if and only if

$$\overline{M} = -\Omega M \Omega \tag{9.71}$$

where M and Ω are now 2×2 matrices, that is, $m = 1$. In an appropriate basis, however, nothing changes; a $2m \times 2m$ complex matrix that satisfies (9.71) can be reinterpreted as an $m \times m$ quaternionic matrix.

If M is complex and satisfies both $M^T \Omega M = \Omega$ and $\overline{M} = -\Omega M \Omega$, then

$$M^\dagger M = \overline{M}^T M = (-\Omega M \Omega)^T M = -\Omega M^T \Omega M = -\Omega \Omega = I \qquad (9.72)$$

and we have shown that

$$\mathrm{SU}(m, \mathbb{H}) \cong \mathrm{Sp}(m). \qquad (9.73)$$

The case $m = 1$ is just

$$\mathrm{Sp}(1) \cong \mathrm{SU}(1, \mathbb{H}) \cong \mathrm{Spin}(3) \qquad (9.74)$$

which is just the isometry group of the quaternions, and, as we have already seen,

$$\mathrm{Sp}(2) \cong \mathrm{SU}(2, \mathbb{H}) \cong \mathrm{Spin}(5) \qquad (9.75)$$

when $m = 2$.

A similar argument can be used to show that $\mathrm{SU}(p, q, \mathbb{H})$ is a symplectic group, often written as $\mathrm{Sp}(p, q, \mathbb{R})$.[7] Thus, *some* symplectic groups can be reinterpreted as quaternionic unitary groups. What about $\mathrm{Sp}(2m, \mathbb{R})$?

Consider $\mathrm{Sp}(4, \mathbb{H})$, which is the double cover of $\mathrm{SO}(3, 2)$. Thus, $\mathrm{Sp}(4, \mathbb{H})$ has $3 \times 2 = 6$ boosts. But it is easy to show that $\mathrm{SU}(p, q, \mathbb{H})$ has $4pq$ boosts. Thus, $\mathrm{Sp}(4, \mathbb{H}) \not\cong \mathrm{SU}(p, q, \mathbb{H})$ for any $p + q = 5$. However, $\mathrm{Sp}(4, \mathbb{H})$ is referred to as the "split" form of $\mathrm{Sp}(2)$, which suggests that we should use the split form of the quaternions, namely $\mathbb{H}'$.

So what is $\mathrm{Sp}(m, \mathbb{H}')$? Again, we start with the case $m = 1$, and again we use Pauli matrices. The split quaternion $a + bL + cK + dKL$ can be represented by

$$a\sigma_t + b\sigma_z - c(i\sigma_y) + d\sigma_x = \begin{pmatrix} a + b & -c + d \\ c + d & a - b \end{pmatrix} \qquad (9.76)$$

where we have written σ_t rather than I for the 2×2 identity matrix. We want to repeat the computation in (9.72), but we must now carefully distinguish between conjugation in $\mathbb{H}'$ and conjugation in $\mathbb{C}$.

Let's start again. According to the Cayley–Dickson process discussed in Section 5.1, we can combine a pair of complex numbers $a, b \in \mathbb{C}$ into a matrix

$$q = \begin{pmatrix} a & -b \\ \overline{b}\epsilon & \overline{a} \end{pmatrix} \qquad (9.77)$$

---

[7]Yet again, there are several conventions for the names of these groups. In particular, $\mathrm{Sp}(m, m, \mathbb{R})$ is *not* the same as $\mathrm{Sp}(m, m)$ as defined implicitly in Section 8.1, nor is $\mathrm{Sp}(2m, \mathbb{R})$ the same as $\mathrm{Sp}(m)$.

where $\epsilon = 1$ if $q \in \mathbb{H}$, and $\epsilon = -1$ if $q \in \mathbb{H}'$. We can generalize this construction to matrices, leading to the representation

$$Q = \begin{pmatrix} A & -B \\ \overline{B}\epsilon & \overline{A} \end{pmatrix} \tag{9.78}$$

of a (possibly split) quaternionic matrix in terms of complex matrices A and B. As in the Cayley–Dickson process, the (quaternionic!) conjugate of Q is obtained by replacing A with $\overline{A}$, and B with $-B$, and the (quaternionic!) transpose of Q is obtained by replacing *each* of A and B with its transpose. We therefore have

$$Q^\dagger = \begin{pmatrix} A^\dagger & B^T \\ -B^\dagger\epsilon & A^T \end{pmatrix} = -\Omega M^T \Omega \tag{9.79}$$

which however only reduces to the *complex* Hermitian conjugate of Q if $\epsilon = 1$. Using quaternionic conjugation throughout, the computation in (9.72) now goes through unchanged whether the components of M are in $\mathbb{H}$ or $\mathbb{H}'$, and we have shown that

$$\mathrm{SU}(m, \mathbb{H}') \cong \mathrm{Sp}(2m, \mathbb{R}). \tag{9.80}$$

Thus, all of the (real forms of the) symplectic groups are indeed quaternionic, but possibly split.

We have just seen that the real symplectic groups are really the quaternionic generalization of the unitary groups. But we can also construct symplectic groups over other division algebras besides the reals.

The definition of symplectic groups given in Section 8.1 is normally used verbatim to define symplectic groups over other division algebras. However, we follow Sudbery [4] in using Hermitian conjugation, rather than transpose, in the definition of generalized symplectic groups. That is, we define

$$\mathrm{Sp}(2m, \mathbb{K}) = \{M \in \mathbb{K}^{2m \times 2m} : M\Omega M^\dagger = \Omega\} \tag{9.81}$$

for $\mathbb{K} = \mathbb{R}, \mathbb{C}, \mathbb{H}, \mathbb{O}$. If $\mathbb{K} = \mathbb{C}$, $i\Omega$ is a Hermitian product of signature (m, m). We can identify these symplectic groups with unitary groups, namely

$$\mathrm{Sp}(2m, \mathbb{C}) \cong \mathrm{SU}(m, m) \tag{9.82}$$

and in particular

$$\mathrm{Sp}(4, \mathbb{C}) \cong \mathrm{SU}(2, 2), \tag{9.83}$$

$$\mathrm{Sp}(6, \mathbb{C}) \cong \mathrm{SU}(3, 3). \tag{9.84}$$

Chapter 10

Lie Groups and Lie Algebras

10.1 Lie Groups

A *Lie group* G, named after the Norwegian mathematician Sophus Lie, is a group whose elements depend smoothly on some number of parameters, and on which the group operations

$$G \times G \longrightarrow G, \tag{10.1}$$
$$(X, Y) \longmapsto X^{-1}Y, \tag{10.2}$$

are smooth. Most of the Lie groups considered here are matrix groups, whose elements are $n \times n$ matrices over some division algebra, and whose group operation is matrix multiplication. The simplest example is SO(2), the rotation group in two dimensions. As discussed in Section 6.2, the elements of SO(2) take the form

$$M(\alpha) = \begin{pmatrix} \cos\alpha & -\sin\alpha \\ \sin\alpha & \cos\alpha \end{pmatrix} \tag{10.3}$$

which clearly depends smoothly on α. Furthermore, we have

$$M(0) = I, \tag{10.4}$$
$$M(\alpha + \beta) = M(\alpha)M(\beta), \tag{10.5}$$

where I denotes the (in this case, 2×2) identity matrix. We refer to

$$\{M(\alpha) : \alpha \in \mathbb{R}\}$$

as a *1-parameter family* of group elements if it satisfies (10.4) and (10.5). We are interested primarily in Lie groups that are *connected to the identity*, in which case every element belongs to at least one 1-parameter family.

The *dimension* $|G|$ of a Lie group G is the number of independent parameters needed to describe it. For example, SO(2) is clearly a

1-dimensional Lie group. What about SO(3)? There are three independent rotations in SO(3), which correctly suggests that the dimension of SO(3) is three. Alternatively, any rotation in three dimensions can be expressed as a product of three rotations about given, fixed axes, for instance using *Euler angles*. More generally, there are $\binom{n}{2} = \frac{n(n-1)}{2}$ independent planes in n dimensions, so that

$$|\mathrm{SO}(n)| = \frac{n(n-1)}{2}. \qquad (10.6)$$

Similar arguments can be used to show that

$$|\mathrm{SU}(n)| = n^2 - 1, \qquad (10.7)$$

$$|\mathrm{Sp}(n)| = n(2n+1). \qquad (10.8)$$

10.2   Lie Algebras

A *Lie algebra*, again named after Sophus Lie, is a vector space $\mathfrak{g}$ together with a binary operation

$$\mathfrak{g} \times \mathfrak{g} \longrightarrow \mathfrak{g}, \qquad (10.9)$$

$$(x, y) \longmapsto [x, y], \qquad (10.10)$$

called the *Lie bracket* of x and y. The Lie bracket is *bilinear* and satisfies

$$[x, y] + [y, x] = 0, \qquad (10.11)$$

$$[x, [y, z]] + [y, [z, x]] + [z, [x, y]] = 0, \qquad (10.12)$$

where the second condition is known as the *Jacobi identity*. Lie algebras can be thought of as infinitesimal Lie groups. More formally, a Lie algebra is the tangent space of the Lie group at the origin, representing infinitesimal "displacements" in all possible "directions" there.

For matrix Lie groups, each 1-parameter family $M(\alpha)$ yields an element A of the corresponding Lie algebra via

$$A = \dot{M} = \left. \frac{\partial M}{\partial \alpha} \right|_{\alpha=0} \qquad (10.13)$$

which is again a matrix. This operation can be reversed through matrix exponentiation (which can be defined formally using power series); we have

$$M(\alpha) = \exp(\dot{M}\alpha). \qquad (10.14)$$

The Lie bracket on a matrix Lie algebra is just the ordinary commutator, that is

$$[x, y] = xy - yx. \qquad (10.15)$$

This operation on matrices clearly satisfies (10.11), and a straightforward computation shows that (10.12) is satisfied as well.

The simplest Lie algebra is $\mathfrak{so}(2)$, the infinitesimal version of the Lie group SO(2), which is the vector space generated by

$$A = \frac{\partial}{\partial \alpha} \begin{pmatrix} \cos\alpha & -\sin\alpha \\ \sin\alpha & \cos\alpha \end{pmatrix} \bigg|_{\alpha=0} = \begin{pmatrix} 0 & -1 \\ 1 & 0 \end{pmatrix}. \tag{10.16}$$

Thus, $\mathfrak{so}(2)$ consists of all real multiplies of A, and is isomorphic to the 1-dimensional vector space $\mathbb{R}$. All commutators on $\mathfrak{so}(2)$ vanish.

The simplest nontrivial Lie algebra is $\mathfrak{so}(3)$, the infinitesimal version of SO(3), which, as shown in Section 7.3, is generated by $(-i$ times) the Pauli matrices. This factor of i represents a major notational difference between two standard notations: Physicists typically insert an i into the differentiation operation (10.13), so that the resulting infinitesimal rotation matrices are *Hermitian*, whereas mathematicians typically omit this i, resulting in *anti-Hermitian* infinitesimal rotation matrices. Adopting this latter convention, $\mathfrak{so}(3)$ is generated by

$$\tau_x = -\frac{i\sigma_x}{2} = \frac{1}{2} \begin{pmatrix} 0 & -i \\ -i & 0 \end{pmatrix}, \tag{10.17}$$

$$\tau_y = -\frac{i\sigma_y}{2} = \frac{1}{2} \begin{pmatrix} 0 & -1 \\ 1 & 0 \end{pmatrix}, \tag{10.18}$$

$$\tau_z = -\frac{i\sigma_z}{2} = \frac{1}{2} \begin{pmatrix} -i & 0 \\ 0 & i \end{pmatrix}, \tag{10.19}$$

which satisfy the *commutation relations*

$$[\tau_x, \tau_y] = \tau_z, \tag{10.20}$$

$$[\tau_y, \tau_z] = \tau_x, \tag{10.21}$$

$$[\tau_z, \tau_x] = \tau_y. \tag{10.22}$$

The *dimension* $|\mathfrak{g}|$ of a Lie algebra $\mathfrak{g}$ is its dimension as a vector space, which is the same as the dimension of the corresponding Lie group. Thus, $\mathfrak{so}(3)$ is a 3-dimensional Lie algebra, and, more generally,

$$|\mathfrak{so}(n)| = \frac{n(n-1)}{2}. \tag{10.23}$$

Furthermore, there are no double cover issues with Lie algebras, so

$$\mathfrak{su}(2) \cong \mathfrak{so}(3) \tag{10.24}$$

without need to involve the spin group Spin(3) (whose Lie algebra is therefore also $\mathfrak{so}(3)$).

10.3   The Classification of Lie Groups

The complete classification of Lie algebras is one of the major results in mathematics, providing a description of the building blocks of all continuous symmetry groups. This classification was first outlined by the German mathematician Wilhelm Killing in the late 1800s, and the first rigorous construction was provided by French mathematician Élie Cartan shortly thereafter; both Killing and Cartan worked with *complex* Lie algebras.

We consider here only *simple Lie algebras*, namely Lie algebras which contain no nontrivial ideals. We also treat the abelian Lie algebra $\mathfrak{so}(2)$ as a special case, which is isomorphic as a vector space to $\mathbb{R}$, and in which all commutators vanish.[1] An *ideal* $\mathfrak{h}$ of a Lie algebra $\mathfrak{g}$ is a subalgebra of $\mathfrak{g}$ with the property that all commutators in $\mathfrak{g}$ involving *one* element of $\mathfrak{h}$ yield another element of $\mathfrak{h}$, that is, for which

$$X \in \mathfrak{h}, Y \in \mathfrak{g} \Longrightarrow [X, Y] \in \mathfrak{h}. \tag{10.25}$$

The zero set $\{0\}$ and the entire Lie algebra $\mathfrak{g}$ are both ideals of $\mathfrak{g}$; a simple Lie algebra has no *other* ideals. Roughly speaking, this means that any proper subalgebra of $\mathfrak{g}$ also mixes up the remaining elements of $\mathfrak{g}$ with each other.

The classification of simple Lie algebras proceeds roughly as follows. First, construct the largest possible subalgebra $\mathfrak{h}$ of $\mathfrak{g}$ such that all elements of $\mathfrak{h}$ commute with each other. There are many ways to do this, but $r = |\mathfrak{h}|$ is the same in each case, and is called the *rank* of the Lie algebra, while $\mathfrak{h}$ itself is called a *Cartan subalgebra* of $\mathfrak{g}$. Now fix a basis of $\mathfrak{h}$, and consider the action of these $|\mathfrak{h}|$ elements on (the rest of) $\mathfrak{g}$. Since our basis elements commute, we can find simultaneous eigenvectors for them (this is why the classification is done over $\mathbb{C}$ rather than $\mathbb{R}$), and in fact we can find a basis of $\mathfrak{g} - \mathfrak{h}$ that consists entirely of simultaneous eigenvectors. The corresponding eigenvalues are vectors in $\mathbb{R}^r$, called the *root system* of $\mathfrak{g}$. The geometry of root systems turns out to be tightly constrained, and can be used to completely classify all simple Lie algebras.

This process sounds harder than it is in practice, so we give an example. We already know from Section 10.2, that a basis for the Lie algebra $\mathfrak{su}(2)$ is $\{\tau_x, \tau_y, \tau_z\}$, none of which commute with each other. Any Cartan subalgebra $\mathfrak{h}$ will therefore be 1-dimensional, so we pick $\mathfrak{h} = \langle \tau_z \rangle$, the 1-dimensional vector space with basis $\{\tau_z\}$. The (complex) eigenvectors of τ_z are

$$[\tau_z, i\tau_x \mp \tau_y] = i\tau_y \pm \tau_x = \mp i(i\tau_x \mp \tau_y). \tag{10.26}$$

---

[1] There is ambiguity in the literature as to whether $\mathfrak{so}(2)$ should be considered "simple".

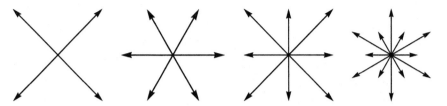

Fig. 10.1 The root diagrams for $\mathfrak{d}_2 = \mathfrak{so}(4)$, $\mathfrak{b}_2 = \mathfrak{so}(5)$, $\mathfrak{a}_2 = \mathfrak{su}(3)$, and $\mathfrak{g}_2$, respectively.

The reader who is familiar with quantum mechanics may recognize

$$L_z = i\hbar\tau_z = \frac{\hbar}{2}\sigma_z \tag{10.27}$$

as the quantum operator corresponding to (the z-component of) spin, and

$$L_\pm = i\hbar\left(\tau_x \mp \tau_y\right) = \frac{\hbar}{2}\left(\sigma_x \pm i\sigma_y\right) \tag{10.28}$$

as the corresponding raising and lowering operators. The classification of simple Lie algebras extends this construction to all (simple) symmetry groups—and works with anti-Hermitian matrices, rather than Hermitian matrices.

The geometric properties of root systems divide them into four infinite families, labeled $\mathfrak{a}_r$, $\mathfrak{b}_r$, $\mathfrak{c}_r$, and $\mathfrak{d}_r$, where r is the rank, as well as five remaining cases, labeled $\mathfrak{g}_2$, $\mathfrak{f}_4$, $\mathfrak{e}_6$, $\mathfrak{e}_7$, and $\mathfrak{e}_8$; these latter five algebras are referred to as the *exceptional Lie algebras*. It turns out that

$$\mathfrak{a}_r \cong \mathfrak{su}(r+1) \qquad (r \geq 1), \tag{10.29}$$

$$\mathfrak{b}_r \cong \mathfrak{so}(2r+1) \qquad (r > 1), \tag{10.30}$$

$$\mathfrak{c}_r \cong \mathfrak{sp}(r) \qquad (r \geq 3), \tag{10.31}$$

$$\mathfrak{d}_r \cong \mathfrak{so}(2r) \qquad (r \geq 4), \tag{10.32}$$

where the restrictions on the rank are to prevent overlaps, since

$$\mathfrak{a}_1 \cong \mathfrak{b}_1 \cong \mathfrak{c}_1, \tag{10.33}$$

$$\mathfrak{b}_2 \cong \mathfrak{c}_2, \tag{10.34}$$

$$\mathfrak{a}_3 \cong \mathfrak{d}_3, \tag{10.35}$$

while $\mathfrak{d}_1 \cong \mathfrak{so}(2) \cong \mathbb{R}$ and $\mathfrak{d}_2 \cong \mathfrak{so}(4) \cong \mathfrak{so}(2) \oplus \mathfrak{so}(2)$ are not simple.

The root diagrams for the rank two Lie algebras $\mathfrak{a}_2$, $\mathfrak{b}_2$, $\mathfrak{d}_2$, and $\mathfrak{g}_2$ are shown in Figure 10.1. Each arrow represents a root; there are $|\mathfrak{g}| - |\mathfrak{h}|$ roots in all, since the Cartan elements are not included. The overall scale is a matter of convention, but the relative scale within each diagram is not. Further discussion of the construction of root diagrams can be found in [6].

Comparing the list of simple Lie algebras and their corresponding Lie groups with our previous results, we see that the $\mathfrak{a}$ family corresponds to the unitary groups $SU(n) = SU(n, \mathbb{C})$, the $\mathfrak{b}$ and $\mathfrak{d}$ families together correspond to the orthogonal groups $SO(n) = SU(n, \mathbb{R})$, and the $\mathfrak{c}$ family corresponds to the symplectic groups $Sp(n) = SU(n, \mathbb{H})$. Do the remaining, exceptional Lie algebras correspond to Lie groups over the octonions? Yes, indeed! We will return to this discussion in Chapter 11.

10.4   Real Forms

A simple Lie algebra admits a non-degenerate inner product, called the *Killing form*. For complex matrix Lie algebras, the Killing form can be taken to be

$$B(X, Y) = \operatorname{tr}(XY). \tag{10.36}$$

A vector space with a non-degenerate inner product admits an orthonormal basis $\{X_m\}$ satisfying

$$B(X_m, X_n) = \pm \delta_{mn} \tag{10.37}$$

where δ_{mn} denotes the Kronecker delta, which is 1 if $m = n$ and 0 otherwise.

Consider the Lie algebra $\mathfrak{so}(3,1)$, corresponding to the Lorentz group $SO(3,1)$ considered in Sections 6.6 and 9.3. Since Lie algebras represent infinitesimal transformations, the question of double covers never arises. For example, at the Lie algebra level,

$$\mathfrak{su}(2) \cong \mathfrak{so}(3), \tag{10.38}$$

$$\mathfrak{sl}(2, \mathbb{C}) \cong \mathfrak{so}(3,1). \tag{10.39}$$

Recalling that the Pauli matrices σ_m each square to the identity matrix, and that the generators of $\mathfrak{su}(2)$ are $\tau_m = -i\sigma_m$, we see that the Killing form is *negative* definite on $\mathfrak{su}(2)$. What about the boosts? The infinitesimal boosts in $\mathfrak{sl}(2, \mathbb{C})$ are just the Pauli matrices themselves! We are therefore led to identify elements with negative squared Killing norm as rotations, and elements with positive squared Killing norm as boosts.

The classification of simple Lie algebras outlined in Section 10.3 involved *complex* Lie algebras. But we have just argued that the complexification of $\mathfrak{su}(2)$ is precisely the complex Lie algebra $\mathfrak{sl}(2, \mathbb{C})$! Put differently, a complex Lie algebra always contains an equal number of boosts and rotations, since you can multiply by i to get from one to the other.

Physical symmetry groups, however, correspond to *real* Lie algebras. Such Lie algebras can be represented using complex matrices; what makes them real is that the *commutators* between elements must all have real coefficients. These coefficients are called *structure constants*. As a *complex* Lie algebra, $\mathfrak{sl}(2, \mathbb{C})$ has just three independent elements—and multiplication by i is allowed. As a *real* Lie algebra, $\mathfrak{sl}(2, \mathbb{C})$ has six independent elements—since multiplication by i is *not* allowed.

From now on, we work exclusively with *real* Lie algebras. As with $\mathfrak{su}(2) \subset \mathfrak{sl}(2, \mathbb{C})$, every real Lie algebra can be regarded as a subalgebra of a complex Lie algebra. But how many real subalgebras does a complex Lie algebra have?

In the case of $\mathfrak{sl}(2, \mathbb{C})$, it is easy to see that we must choose either σ_x or τ_x. If we have both, then either there are no other elements, in which case the algebra is abelian (and not simple), or it includes all of $\mathfrak{sl}(2, \mathbb{C})$. Similar considerations apply to the y and z basis elements. It therefore appears that we have $2^3 = 8$ possible 3-dimensional real subalgebras. However, remember that the structure constants must be real; the algebra must close, with real coefficients. Due to the cyclic symmetry among x, y, and z, there are only two inequivalent 3-dimensional real subalgebras of $\mathfrak{sl}(2, \mathbb{C})$, generated either by $\{\tau_x, \tau_y, \tau_z\}$ or by $\{\sigma_x, \sigma_y, \tau_z\}$. We say that there are two real forms of $\mathfrak{a}_1 = \mathfrak{su}(2)$, namely $\mathfrak{su}(2)$ (all rotations), and $\mathfrak{su}(1, 1) = \mathfrak{so}(2, 1)$ (two boosts and one rotation). Since rotations have compact orbits (and boosts do not), the Lie groups corresponding to real forms containing only rotations are themselves compact. We use the same language for Lie groups: The two real forms of SO(3) are SO(3) itself, which is compact, and SO(2, 1), which is not.

This idea carries over to real forms of other Lie algebras. Different real forms of a given Lie algebra must all have the same complexification, that is, they are different real subalgebras of a given complex Lie algebra. In most cases, each such real form is fully determined by giving its Killing–Cartan classification and the number of boosts it contains. Determining the allowed number of boosts requires checking under what circumstances the structure constants are real. We already know some examples: Each $\mathfrak{so}(p, q)$ is a real form of $\mathfrak{so}(p + q)$ (namely the one with pq boosts), and each $\mathfrak{su}(p, q)$ is a real form of $\mathfrak{su}(p + q)$ (namely the one with $2pq$ boosts). But these are not the only real forms.

Chapter 11

The Exceptional Groups

11.1 The Geometry of G_2

What are the symmetries of the division algebras $\mathbb{R}$, $\mathbb{C}$, $\mathbb{H}$, and $\mathbb{O}$? Let's start with $\mathbb{H}$. The labeling i, j, k is arbitrary; *any* right-handed, orthonormal basis of $\operatorname{Im}\mathbb{O}$ could be used instead of i, j, k. This freedom represents a *geometric* symmetry, corresponding to the rigid rotations on the unit sphere in the 3-dimensional vector space of imaginary quaternions. But this symmetry is also *algebraic*, since the choice of handedness ensures that the multiplication table is preserved.

More formally, we define the *isometry group* $\operatorname{SO}(\operatorname{Im}\mathbb{K})$ to be the group of orientation- and norm-preserving transformations on the (normed) vector space $\operatorname{Im}\mathbb{K}$ of imaginary elements of $\mathbb{K}$, and we define the *automorphism group* $\operatorname{Aut}(\mathbb{K})$ to be the transformations Φ on $\mathbb{K}$ that preserve the multiplication table, that is, that satisfy

$$\Phi(pq) = \Phi(p)\,\Phi(q). \tag{11.1}$$

Any automorphism must preserve the identity element, that is,

$$\Phi(1) = 1 \tag{11.2}$$

so that $\operatorname{Aut}(\mathbb{K})$ acts only on $\operatorname{Im}\mathbb{K}$. Over the quaternions, we have

$$\operatorname{Aut}(\mathbb{H}) = \operatorname{SO}(\operatorname{Im}\mathbb{H}) = \operatorname{SO}(3). \tag{11.3}$$

What are the symmetries of $\mathbb{C}$? The imaginary complex numbers $\operatorname{Im}\mathbb{C}$ form a 1-dimensional vector space, whose only symmetry is the orientation-reversing transformation $i \longmapsto -i$. This transformation is in fact an automorphism, but such discrete transformations do not generate 1-parameter families. And $\mathbb{R}$ contain no imaginary units; its isometry and automorphism groups are trivial.

What about the octonions? There are seven basis imaginary units, so the isometry group is clearly

$$SO(\text{Im}\,\mathbb{O}) = SO(7). \tag{11.4}$$

However, a simple counting argument shows that not all of the 21 elements of $SO(7)$ can be automorphisms.

Any automorphism of $\mathbb{O}$ must take i somewhere. Since the imaginary units form a 6-sphere in $\text{Im}\,\mathbb{O}$, there are six degrees of freedom in where i can go. Next, we place j. Where can it go? Somewhere on the 6-sphere, but orthogonal to i, corresponding to an additional five degrees of freedom. However, once i and j have been placed, we have no choice in where we put k, since $\Phi(k) = \Phi(i)\Phi(j)$. Having placed i, j, k, we must put ℓ on the 6-sphere so that it is orthogonal to each of them, corresponding to $6 - 3 = 3$ additional degrees of freedom. The remaining basis units, $i\ell$, $j\ell$, $k\ell$, are again constrained by the automorphism property. Thus, there are $6 + 5 + 3 = 14$ choices we can make, and we expect

$$|\text{Aut}(\mathbb{O})| = 14. \tag{11.5}$$

The automorphism group $\text{Aut}(\mathbb{O})$ is the smallest of the exceptional Lie groups, and is classified as G_2, that is

$$G_2 = \text{Aut}(\mathbb{O}). \tag{11.6}$$

But what are the 14 independent automorphisms of $\mathbb{O}$?

Consider a rotation in the ij-plane. We know this transformation is an automorphism of $\mathbb{H}$, so the product of the rotated units is still k.[1] But we have also affected products such as $i(i\ell)$—unless we also rotate by an equal amount in the $(i\ell)(j\ell)$-plane. Direct calculation shows that this is sufficient; the transformation

$$i \longmapsto i\cos\alpha - j\sin\alpha,$$
$$j \longmapsto i\sin\alpha + j\cos\alpha,$$
$$k \longmapsto k,$$
$$k\ell \longmapsto k\ell,$$
$$j\ell \longmapsto i\ell\sin\alpha + j\ell\cos\alpha,$$
$$i\ell \longmapsto i\ell\cos\alpha - j\ell\sin\alpha,$$
$$\ell \longmapsto \ell, \tag{11.7}$$

---

[1]This statement is equivalent to the more familiar statement that the cross product of *any* two (oriented) unit vectors in the xy-plane gives the unit vector in the z-direction.

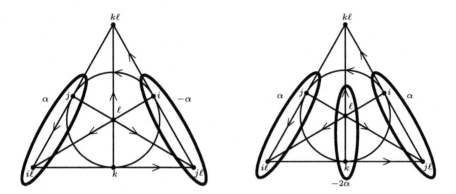

Fig. 11.1   Two independent elements of G_2 that "point" to $k\ell$.

is an automorphism of $\mathbb{O}$ for each value of α.

How many such (families of) automorphisms are there? We rotated two planes, but there is a relationship between the planes. We say that the ij-plane "points" to $ij = k$. Similarly, the $(j\ell)(i\ell)$ plane also points to k. Thus, a better description of our automorphism is that it involves *equal but opposite rotations in two planes that point to the same unit*. There are seven units that can be pointed to, and in each case three choices of a pair of planes pointing to that unit. We therefore have 21 automorphisms of this form, but not all of them are independent.

Which planes point to k? Which basis units multiply to k? There are three lines intersecting at each point in the multiplication table of $\mathbb{O}$, represented in terms of the 7-point projective plane, so there are three such pairs of units, in this case

$$k = ij = (j\ell)(i\ell) = \ell(k\ell). \tag{11.8}$$

An element of G_2 "pointing" to $k\ell$ is illustrated in the first diagram in Figure 11.1. However, combining any two pairs of rotations (by the same angle) yields the third pair, so only two of these three automorphisms are independent. Thus, we have precisely $7 \times 2 = 14$ independent automorphisms, which generate G_2. But which 14 automorphisms should we pick to be our basis?

We adopt the following strategy. Choose one octonion to be special, say ℓ. At each of the six points other than ℓ in the projective plane, three lines intersect. One of these lines contains ℓ, and the other two do not. Thus, there is a preferred pair of planes pointing to each unit other than ℓ. For example, our construction above started by rotating the ij- and $(j\ell)(i\ell)$-planes by equal and opposite amounts—and leaving the $\ell(k\ell)$-plane alone.

Call this automorphism A_k. Rather than picking one of the other two automorphisms pointing to k as special, take their difference, and call it G_k. With a particular choice of orientation, G_k will rotate the ij- and $(j\ell)(i\ell)$-planes by the *same* amount—and will rotate the $\ell(k\ell)$-plane by twice as much in the opposite direction. Repeat this construction for automorphisms pointing to the other units, and make an arbitrary choice of which pair of planes to use when constructing A_ℓ. The 14 transformations $\{A_m, G_m\}$ generate G_2. The G_2 elements $A_{k\ell}$ and $G_{k\ell}$ are illustrated in Figure 11.1.

We are not quite done. There is a third independent rotation, call it S_m, pointing to each unit m, namely the transformation that rotates all three planes in the same direction. For example, S_k rotates the ij-, $(j\ell)(i\ell)$-, and $\ell(k\ell)$-planes by equal amounts in the same direction. The rotations S_m are of course *not* in G_2, but they are precisely the seven extra rotations needed to extend G_2 to SO(7).

Having treated ℓ as special, it is of interest to know what subgroups of G_2 and SO(7) leave ℓ invariant. We constructed the A_m precisely so that they would fix ℓ, so the subgroup of G_2 that fixes ℓ is generated by the seven transformations A_m together with G_ℓ—any rotation in planes pointing to ℓ must fix ℓ. These eight transformations turn out to generate a copy of SU(3). The subgroup of SO(7) that fixes one of the units is of course SO(6), which must contain this SU(3). Since $|SU(3)| = 8$ and $|SO(6)| = 15$, we are missing seven independent elements of SO(6). Which ones? The rotations that rotate two planes pointing to m in the *same* direction, leaving the third alone. For example, pointing to k, the missing element of SO(6) can be chosen to be the rotation rotating the ij- and $(j\ell)(i\ell)$-planes by the same amount in the same direction.

11.2   The Albert Algebra

In order to discuss the remaining four exceptional Lie groups, we introduce the algebra of 3×3 octonionic Hermitian matrices, known as the *Albert algebra*, written as $\mathbf{H}_3(\mathbb{O})$. Since the product of Hermitian matrices is not necessarily Hermitian, we introduce the *Jordan product*

$$\mathcal{A} \circ \mathcal{B} = \frac{1}{2}(\mathcal{A}\mathcal{B} + \mathcal{B}\mathcal{A}) \tag{11.9}$$

for elements $\mathcal{A}, \mathcal{B} \in \mathbf{H}_3(\mathbb{O})$, which we henceforth refer to as *Jordan matrices*. It is easily checked that the Jordan product takes Hermitian matrices to Hermitian matrices, is (obviously) commutative, but is not associative.

A *Jordan algebra* is an algebra with a commutative product ∘ that also satisfies

$$(\mathcal{A} \circ \mathcal{B}) \circ \mathcal{A}^2 = \mathcal{A} \circ (\mathcal{B} \circ \mathcal{A}^2) \tag{11.10}$$

where of course

$$\mathcal{A}^2 = \mathcal{A} \circ \mathcal{A}. \tag{11.11}$$

Jordan algebras were introduced by the German mathematical physicist Pascual Jordan in order to represent observables in quantum mechanics. All Jordan algebras *except* the Albert algebra arise by introducing the Jordan product on an *associative* algebra; the Albert algebra is therefore also referred to as the *exceptional Jordan algebra*. Higher powers must be explicitly defined, such as

$$\mathcal{A}^3 = \mathcal{A}^2 \circ \mathcal{A} = \mathcal{A} \circ \mathcal{A}^2. \tag{11.12}$$

We also introduce the *Freudenthal product* of two Jordan matrices $\mathcal{A}$ and $\mathcal{B}$, given by

$$\mathcal{A} * \mathcal{B} = \mathcal{A} \circ \mathcal{B} - \frac{1}{2}\Big(\mathcal{A}\,\mathrm{tr}\,(\mathcal{B}) + \mathcal{B}\,\mathrm{tr}\,(\mathcal{A})\Big) \tag{11.13}$$

$$+ \frac{1}{2}\Big(\mathrm{tr}\,(\mathcal{A})\,\mathrm{tr}\,(\mathcal{B}) - \mathrm{tr}\,(\mathcal{A} \circ \mathcal{B})\Big) \tag{11.14}$$

where the identity matrix is implicit in the last term. The determinant of a Jordan matrix can then be defined as

$$\det(\mathcal{A}) = \frac{1}{3}\,\mathrm{tr}\,\Big((\mathcal{A} * \mathcal{A}) \circ \mathcal{A}\Big). \tag{11.15}$$

Concretely, if

$$\mathcal{A} = \begin{pmatrix} p & \bar{a} & c \\ a & m & \bar{b} \\ \bar{c} & b & n \end{pmatrix} \tag{11.16}$$

with $p, m, n \in \mathbb{R}$ and $a, b, c \in \mathbb{O}$ then

$$\det \mathcal{A} = pmn + c(ba) + \overline{c(ba)} - n|a|^2 - p|b|^2 - m|c|^2. \tag{11.17}$$

These products become somewhat less mysterious if we consider the restriction of $\mathbb{O}$ to $\mathbb{R}$, and suppose that

$$\mathcal{A} = vv^{\dagger}, \qquad \mathcal{B} = ww^{\dagger}, \tag{11.18}$$

where $v, w \in \mathbb{R}^3$ (and of course Hermitian conjugation, denoted by †, reduces to the matrix transpose). In this case,

$$\mathrm{tr}\,(\mathcal{A} \circ \mathcal{B}) = (v \cdot w)^2, \tag{11.19}$$

$$\mathcal{A} * \mathcal{B} = (v \times w)(v \times w)^{\dagger}, \tag{11.20}$$

where · and × denote the ordinary dot and cross product. The Jordan product ∘ and the Freudenthal product * can therefore be thought of as generalizations of the dot and cross products, respectively.

11.3   The Geometry of F_4

As suggested in Section 11.2, the Albert algebra $\mathbf{H}_3(\mathbb{O})$ can be regarded as a generalization of the vector space $\mathbb{R}^3$ to the octonions. The elements $\mathcal{X} \in \mathbf{H}_3(\mathbb{O})$ of the Albert algebra are Hermitian matrices, and have well-defined determinants. We therefore seek transformations that preserve these properties.

Reasoning by analogy with Section 7.3, we seek transformations of the form

$$\mathcal{X} \longmapsto \mathcal{M}\mathcal{X}\mathcal{M}^\dagger \tag{11.21}$$

that preserve the determinant and trace of $\mathcal{X}$; this group would deserve the name SU$(3, \mathbb{O})$.

Before proceeding further, however, we must ask under what circumstances $\mathcal{M}\mathcal{X}\mathcal{M}^\dagger$ is well-defined, that is, when is it true that

$$(\mathcal{M}\mathcal{X})\mathcal{M}^\dagger = \mathcal{M}(\mathcal{X}\mathcal{M}^\dagger). \tag{11.22}$$

As in Section 9.3, we therefore assume that the elements of $\mathcal{M}$ lie in some complex subalgebra of $\mathbb{O}$, which can be different for different $\mathcal{M}$, and we will need to allow *nesting*, that is, not all transformations can be carried out with a single $\mathcal{M}$.

In Section 9.2, we showed that SU$(2, \mathbb{O}) \cong$ SO(9). There are three obvious ways to embed 2×2 matrices inside 3×3 matrices, depending on which row and column are ignored; we refer to these alternatives as *types*. Thus, SU$(3, \mathbb{O})$ is contained in the union of the three copies of SU$(2, \mathbb{O})$, each of which has 36 elements. However, there is substantial overlap between these three subgroups.

Recall from Section 9.1.4 that left, right, and symmetric multiplication over $\mathbb{O}$ all yield representations of SO(8), related by triality. Consider now the three independent off-diagonal elements of a Jordan matrix $\mathcal{X}$, each of which is of course an element of $\mathbb{O}$. Consider further the *diagonal* elements of SU$(2, \mathbb{O})$, embedded (in three ways) in SU$(3, \mathbb{O})$, acting on $\mathcal{X}$; such elements take the form[2]

$$\mathcal{M} = \begin{pmatrix} p & 0 & 0 \\ 0 & p & 0 \\ 0 & 0 & 1 \end{pmatrix} \tag{11.23}$$

---

[2]Elements of type (11.23) must in general be nested.

with $p^2 = -1$, or

$$\mathcal{M} = \begin{pmatrix} q & 0 & 0 \\ 0 & \bar{q} & 0 \\ 0 & 0 & 1 \end{pmatrix} \tag{11.24}$$

where $|q| = 1$, or cyclic permutations of these matrices. Under (11.21), each such transformation leaves the diagonal of $\mathcal{X}$ alone, and acts separately on the three off-diagonal elements, in each case implementing one of the three representations of SO(8).

We conclude that the three copies of SO(8), coming from the three copies of SU(2, $\mathbb{O}$) embedded in SU(3, $\mathbb{O}$), must in fact be the same, which is an important consequence of triality. Redoing our count, we now have a *single* copy of SO(8), with its 28 elements. However, the eight remaining transformations coming from each copy of SU(2, $\mathbb{O}$) are indeed different (since they act on different pairs of diagonal elements), so we have $28 + 3 \times 8 = 52$ elements in all, that is,

$$|\mathrm{SU}(3, \mathbb{O})| = 52. \tag{11.25}$$

This group is one of the exceptional Lie groups, and is classified as F_4, that is,

$$F_4 = \mathrm{SU}(3, \mathbb{O}). \tag{11.26}$$

We digress briefly to discuss some further properties of triality, which play an essential role in the above description of F_4. Including cyclic permutations, there are only 14 independent transformations of the form (11.24), since the third set of seven can be constructed from the other two. These 14 transformations act on the three octonions in $\mathcal{X}$, acting on one octonion by symmetric multiplication, another by left multiplication, and the third by right multiplication. Each of these transformations is in SO(8), but they are different from each other, related by triality. As for the remaining transformations, of the form (11.23), there appear to be only seven of them, but there are really $\binom{7}{2} = 21$, due to nesting. However, seven of these can also be represented using (cyclically permuted) transformations of the form (11.24); the remaining 14 transformations, which can only be represented using nesting, are precisely the generators of G_2. In this case, and only in this case, symmetric, left, and right multiplication yield the *same* transformation in SO(8), a property that we refer to as *strong triality*. An example of strong triality is the identity

$$-i(j(kxk)j)i = -i(j(kx)) = ((xk)j)i \tag{11.27}$$

for any $x \in \mathbb{O}$ (where the minus signs are needed due to the odd number of factors in this nonstandard instance of nesting).

We conclude this section by listing an explicit set of 52 generators for F_4.

- There are seven independent elements of the form (11.23), which when nested yield the $\binom{7}{2} = 14$ generators of G_2. Since a rotation in a single plane is represented by two nested flips, a typical G_2 transformation involving two such rotations requires *four* nested flips. An example of such a G_2 transformation is

$$\mathcal{X} \longmapsto \mathcal{M}_4 \left(\mathcal{M}_3 \left(\mathcal{M}_2 \left(\mathcal{M}_1 \mathcal{X} \mathcal{M}_1^\dagger \right) \mathcal{M}_2^\dagger \right) \mathcal{M}_3^\dagger \right) \mathcal{M}_4^\dagger \qquad (11.28)$$

where

$$\mathcal{M}_1 = \begin{pmatrix} i & 0 & 0 \\ 0 & i & 0 \\ 0 & 0 & 1 \end{pmatrix}, \qquad (11.29)$$

$$\mathcal{M}_2 = \begin{pmatrix} p & 0 & 0 \\ 0 & p & 0 \\ 0 & 0 & 1 \end{pmatrix}, \qquad (11.30)$$

$$\mathcal{M}_3 = \begin{pmatrix} i\ell & 0 & 0 \\ 0 & i\ell & 0 \\ 0 & 0 & 1 \end{pmatrix}, \qquad (11.31)$$

$$\mathcal{M}_4 = \begin{pmatrix} q & 0 & 0 \\ 0 & q & 0 \\ 0 & 0 & 1 \end{pmatrix}, \qquad (11.32)$$

with $p = i \cos\alpha + j \sin\alpha$ and $q = i\ell \cos\alpha + j\ell \sin\alpha$. In Section 11.1, we referred to this transformation (acting on $\mathbb{O}$) as A_k.

- There are seven elements of SO(8) of the form

$$\mathcal{M} = \begin{pmatrix} e^{i\alpha} & 0 & 0 \\ 0 & e^{i\alpha} & 0 \\ 0 & 0 & e^{-2i\alpha} \end{pmatrix} \qquad (11.33)$$

again acting via (11.21). We will refer to this transformation as S_i. These transformations implement "type I" SO(7) $\subset$ SO(8) transformations, so called because of their block structure.

- The remaining seven independent elements of SO(8) have the form (11.24). An example of such an SO(8) transformation is

$$\mathcal{M} = \begin{pmatrix} e^{i\alpha} & 0 & 0 \\ 0 & e^{-i\alpha} & 0 \\ 0 & 0 & 1 \end{pmatrix} \qquad (11.34)$$

acting via (11.21). We will refer to this transformation as D_i.

- There are an additional eight elements of "type I" SO(9), which take the form

$$\mathcal{M} = \begin{pmatrix} \cos\alpha & -\overline{q}\sin\alpha & 0 \\ q\sin\alpha & \cos\alpha & 0 \\ 0 & 0 & 1 \end{pmatrix} \tag{11.35}$$

where $q = \{1, i, j, k, k\ell, j\ell, i\ell, \ell\}$, together with $2 \times 8 = 16$ further elements obtained from these by cyclic permutations ("types II and III"). We will refer to these transformations as R_q^I, R_q^{II}, and R_q^{III}, respectively.

We have exhibited an explicit set of $14 + 7 + 7 + 3 \times 8 = 52$ generators for F_4, as claimed.

11.4 The Geometry of E_6

In Section 11.3, we discussed the transformations that preserve the determinant and trace of Jordan matrices $\mathcal{X} \in \mathbf{H}_3(\mathbb{O})$; such transformations are elements the group $F_4 = \mathrm{SU}(3, \mathbb{O})$. It is now straightforward to remove the restriction on the trace, and consider all transformations on $\mathbf{H}_3(\mathbb{O})$ that preserve the determinant; this group would deserve the name $\mathrm{SL}(3, \mathbb{O})$.

Again, we make use of our knowledge of the 2×2 case, in this case the transition from $\mathrm{SU}(2, \mathbb{O}) \cong \mathrm{Spin}(9)$ to $\mathrm{SL}(2, \mathbb{O}) \cong \mathrm{Spin}(9, 1)$, as discussed in Sections 9.2 and 9.3. This transition involves the addition of nine boosts, and there are three natural embeddings of 2×2 matrices inside 3×3 matrices ("types"), so we expect to add $3 \times 9 = 27$ boosts to $\mathrm{SU}(3, \mathbb{O})$ in order to get $\mathrm{SL}(3, \mathbb{O})$. However, three of these boosts are diagonal matrices, namely the boosts with "z", and only two of these are independent. Thus, $\mathrm{SL}(3, \mathbb{O})$ contains $52 + 26 = 78$ elements, that is,

$$|\mathrm{SL}(3, \mathbb{O})| = 78. \tag{11.36}$$

This group is another of the exceptional Lie groups, and is classified as E_6. More precisely, $\mathrm{SL}(3, \mathbb{O})$ is a particular real form of E_6, namely the one with 52 rotations and 26 boosts; its *signature* is therefore $26 - 52 = -26$, and we write

$$E_{6(-26)} = \mathrm{SL}(3, \mathbb{O}). \tag{11.37}$$

For completeness, we list the 78 elements of $\mathrm{SL}(3, \mathbb{O})$ here explicitly. Since

$$\mathrm{SU}(3) \subset G_2 \subset \mathrm{SO}(7) \subset \mathrm{SO}(8) \subset \mathrm{SU}(2, \mathbb{O}) \subset F_4 \subset E_{6(-26)}, \tag{11.38}$$

all of these groups are subgroups of $SL(3, \mathbb{O})$, as is $SL(2, \mathbb{O})$. The elements of these subgroups are also identified below.

- As discussed in Section 11.1, the $SU(3) \subset G_2$ that fixes ℓ is generated by the eight elements $\{A_i,\, A_j,\, A_k,\, A_{k\ell},\, A_{j\ell},\, A_{i\ell},\, A_\ell,\, G_\ell\}$.
- The six remaining generators of G_2 are $\{G_i,\, G_j,\, G_k,\, G_{k\ell},\, G_{j\ell},\, G_{i\ell}\}$.[3]
- The seven remaining generators of (type I) $SO(7)$ are $\{S_i,\, S_j,\, S_k,\, S_{k\ell},\, S_{j\ell},\, S_{i\ell},\, S_\ell\}$.
- The seven remaining generators of $SO(8)$ are $\{D_i,\, D_j,\, D_k,\, D_{k\ell},\, D_{j\ell},\, D_{i\ell},\, D_\ell\}$.
- The eight remaining generators of $SU(2, \mathbb{O}) \cong SO(9)$ are $\{R_1^I,\, R_i^I,\, R_j^I,\, R_k^I,\, R_{k\ell}^I,\, R_{j\ell}^I,\, R_{i\ell}^I,\, R_\ell^I\}$.
- The 16 remaining generators of F_4 are $\{R_1^{II},\, R_i^{II},\, R_j^{II},\, R_k^{II},\, R_{k\ell}^{II},\, R_{j\ell}^{II},\, R_{i\ell}^{II},\, R_\ell^{II}\}$ and $\{R_1^{III},\, R_i^{III},\, R_j^{III},\, R_k^{III},\, R_{k\ell}^{III},\, R_{j\ell}^{III},\, R_{i\ell}^{III},\, R_\ell^{III}\}$.
- The nine boosts needed to get from $SU(2, \mathbb{O})$ to $SL(2, \mathbb{O}) \cong SO(9,1)$ were given (as 2×2 matrices) in Section 9.3. Rewritten as 3×3 matrices, they take the form

$$B_q^I = \begin{pmatrix} \cosh\alpha & \bar{q}\sinh\alpha & 0 \\ q\sinh\alpha & \cosh\alpha & 0 \\ 0 & 0 & 1 \end{pmatrix} \qquad (11.39)$$

for $q = \{1, i, j, k, k\ell, j\ell, i\ell, \ell\}$, as well as

$$B_z^I = \begin{pmatrix} e^\alpha & 0 & 0 \\ 0 & e^{-\alpha} & 0 \\ 0 & 0 & 1 \end{pmatrix}. \qquad (11.40)$$

- Finally, the remaining 17 generators of E_6 are the 16 boosts B_q^{II} and B_q^{III} obtained by cyclically permuting (11.39), together with the single independent boost obtained by cyclically permuting (11.40), which we take to be

$$B_3 = \begin{pmatrix} e^\alpha & 0 & 0 \\ 0 & e^\alpha & 0 \\ 0 & 0 & e^{-2\alpha} \end{pmatrix}. \qquad (11.41)$$

We have exhibited an explicit set of $8 + 6 + 7 + 7 + 8 + 16 + 9 + 17 = 78$ generators for E_6, as claimed.

---

[3]The generators of G_2 given here differ slightly from those given in Section 11.3.

11.5 The Geometry of E_7[4]

Freudenthal [8] provided a description of the Lie algebra $\mathfrak{e}_{7(-25)}$, denoted throughout this section simply as $\mathfrak{e}_7$. Barton & Sudbery [9] showed algebraically how to interpret this Lie algebra as the symplectic algebra $\mathfrak{sp}(6, \mathbb{O})$. Here we discuss the geometric interpretation of this symplectic algebra [7], providing a natural symplectic interpretation of its minimal representation. We work throughout with the Lie algebra, but, given our explicit matrix description of $E_{6(-26)}$ in terms of nested operations, it is not difficult to reinterpret our discussion at the group level. In particular, since a Lie algebra is just an infinitesimal representation of a Lie group, both act on the same vector spaces. Our geometric description of the vector space being acted on therefore applies equally well in both cases, even though the objects (nested matrices) acting on it differ.

Let $\mathcal{X}, \mathcal{Y} \in \mathbf{H}_3(\mathbb{O})$ be elements of the Albert algebra, that is, 3×3 Hermitian matrices whose components are octonions. As discussed at the group level in Section 11.4, the Lie algebra $\mathfrak{e}_{6(-26)}$ (henceforth denoted $\mathfrak{e}_6$) acts on the Albert algebra $\mathbf{H}_3(\mathbb{O})$. The generators of $\mathfrak{e}_6$ fall into one of three categories; there are 26 *boosts*, 14 *derivations*,[5] and 38 other *rotations* (the remaining generators of $\mathfrak{f}_4$). For both boosts and rotations, $\phi \in \mathfrak{e}_6$ can be treated as a 3×3, tracefree, octonionic matrix; boosts are Hermitian, and rotations are anti-Hermitian. Such matrices $\phi \in \mathfrak{e}_6$ act on the Albert algebra via[6]

$$\mathcal{X} \longmapsto \phi \mathcal{X} + \mathcal{X} \phi^\dagger \tag{11.42}$$

where $\dagger$ denotes conjugate transpose (in $\mathbb{O}$). As discussed at the group level for automorphisms in Sections 11.1 and 11.3, the derivations can be obtained by successive rotations (or boosts) through *nesting*, corresponding to commutators in the Lie algebra. It therefore suffices to consider the boosts and rotations, that is, to consider matrix transformations.[7]

---

[4]The material in this section is adapted from [7].

[5]Derivations are infinitesimal automorphisms, so from (11.1) we see that the derivations of $\mathbb{K}$ satisfy $\phi(pq) = \phi(p)\, q + p\, \phi(q)$ for all $p, q \in \mathbb{K}$. The derivations of $\mathbb{O}$ are the elements of $\mathfrak{g}_2$.

[6]This action is the infinitesimal version of the group action $\mathcal{X} \longmapsto \mathcal{M}\mathcal{X}\mathcal{M}^\dagger$. As an element of a Lie group, $\mathcal{M}$ can be assumed to be part of a 1-parameter family $\mathcal{M}(\lambda)$, with $\mathcal{M}(0) = \mathcal{I}$. Then ϕ is just $\frac{d\mathcal{M}}{d\lambda}\big|_{\lambda=0}$, and (11.42) follows from the product rule.

[7]Since all rotations can be obtained from pairs of boosts, it would be enough to consider boosts alone.

The dual representation of $\mathfrak{e}_6$ is formed by the duals ϕ' of each $\phi \in \mathfrak{e}_6$, defined via

$$\mathrm{tr}\left(\phi(\mathcal{X}) \circ \mathcal{Y}\right) = -\mathrm{tr}\left(\mathcal{X} \circ \phi'(\mathcal{Y})\right) \tag{11.43}$$

for $\mathcal{X}, \mathcal{Y} \in \mathbf{H}_3(\mathbb{O})$. It is easily checked that $\phi' = \phi$ on rotations, but that $\phi' = -\phi$ on boosts. Thus,

$$\phi' = -\phi^\dagger \tag{11.44}$$

for both boosts and rotations.

We can regard $\mathfrak{e}_7$ as the conformal algebra associated with $\mathfrak{e}_6$, since $\mathfrak{e}_7$ consists of the 78 elements of $\mathfrak{e}_6$, together with 27 translations, 27 conformal translations, and a dilation. We follow Freudenthal in representing elements of $\mathfrak{e}_7$ as

$$\Theta = (\phi, \rho, \mathcal{A}, \mathcal{B}) \tag{11.45}$$

where $\phi \in \mathfrak{e}_6$, $\rho \in \mathbb{R}$ is the dilation, and $\mathcal{A}, \mathcal{B} \in \mathbf{H}_3(\mathbb{O})$ are elements of the Albert algebra, representing the translations.[8]

What does Θ act on? We again follow Freudenthal, who describes the elements of the minimal representation of $\mathfrak{e}_7$ in the form

$$\mathcal{P} = (\mathcal{X}, \mathcal{Y}, p, q) \tag{11.46}$$

where $\mathcal{X}, \mathcal{Y} \in \mathbf{H}_3(\mathbb{O})$, and $p, q \in \mathbb{R}$, and then gives the action of Θ on $\mathcal{P}$ as

$$\mathcal{X} \longmapsto \phi(\mathcal{X}) + \frac{1}{3} \rho \mathcal{X} + 2\mathcal{B} * \mathcal{Y} + \mathcal{A} q, \tag{11.47}$$

$$\mathcal{Y} \longmapsto 2\mathcal{A} * \mathcal{X} + \phi'(\mathcal{Y}) - \frac{1}{3} \rho \mathcal{Y} + \mathcal{B} p, \tag{11.48}$$

$$p \longmapsto \mathrm{tr}\left(\mathcal{A} \circ \mathcal{Y}\right) - \rho p, \tag{11.49}$$

$$q \longmapsto \mathrm{tr}\left(\mathcal{B} \circ \mathcal{X}\right) + \rho q. \tag{11.50}$$

But how are we to visualize this action?

11.5.1 *The Symplectic Structure of $\mathfrak{so}(k+2, 2)$*

Consider first the analogous problem using 2×2 matrices [10–12]; see also Section 15.1. Elements of $\mathfrak{so}(k + 2, 2)$, where $k = |\mathbb{K}| = 1, 2, 4, 8$, can be represented in terms of actions on 4×4 matrices of the form

$$P_0 = \begin{pmatrix} p\,\mathbf{I} & \mathbf{X} \\ -\widetilde{\mathbf{X}} & q\,\mathbf{I} \end{pmatrix} \tag{11.51}$$

---

[8]The translations and conformal translations are sums and differences of $\mathcal{A}$ and $\mathcal{B}$, which individually turn out be *null*, that is, neither boosts nor rotations.

where X is a 2×2 Hermitian matrix over $\mathbb{K}$ carrying a representation of $\mathfrak{so}(k+1,1)$ (see Section 9.3), $p, q \in \mathbb{R}$, I denotes the 2×2 identity matrix, and tilde denotes trace-reversal, that is, $\widetilde{X} = X - \operatorname{tr}(X)\,I$. The matrix P_0 can be thought of as the upper right 4×4 block of an 8×8 Clifford algebra representation over $\mathbb{K}$, and the action of $\mathfrak{so}(k+2,2)$ on P_0 is obtained as usual from (the restriction of) the quadratic elements of the Clifford algebra. The generators $A \in \mathfrak{so}(k+2,2)$ can be chosen so that the action takes the form

$$P_0 \longmapsto AP_0 \pm P_0 A \qquad (11.52)$$

where the case-dependent signs are related to the restriction from 8×8 matrices to 4×4 matrices. Following Sudbery [4], we define the elements A of the symplectic Lie algebra $\mathfrak{sp}(4, \mathbb{K})$ by the condition[9]

$$A\Omega + \Omega A^\dagger = 0 \qquad (11.53)$$

where

$$\Omega = \begin{pmatrix} 0 & I \\ -I & 0 \end{pmatrix}. \qquad (11.54)$$

Solutions of (11.53) take the form

$$A = \begin{pmatrix} \phi - \frac{1}{2}\rho\,I & A \\ B & -\phi^\dagger + \frac{1}{2}\rho\,I \end{pmatrix} \qquad (11.55)$$

where both A and B are Hermitian, $\operatorname{tr}(\phi) = 0$, and $\rho \in \mathbb{R}$. But generators of $\mathfrak{so}(k+2,2)$ take exactly the same form: ϕ represents an element of $\mathfrak{so}(k+1,1)$, A and B are (null) translations, and ρ is the dilation. Direct computation shows that the generators A of $\mathfrak{so}(k+2,2)$ do indeed satisfy (11.53), so that $\mathfrak{so}(k+2,2)$ and $\mathfrak{sp}(4,\mathbb{K})$ can be identified as vector spaces, and hence also as Lie algebras; the above construction therefore establishes the isomorphism

$$\mathfrak{so}(k+2,2) \cong \mathfrak{sp}(4,\mathbb{K}). \qquad (11.56)$$

We can bring the representation (11.51) into a more explicitly symplectic form by treating X as a vector-valued 1-form, and computing its Hodge dual $*X$, defined by

$$*X = X\epsilon \qquad (11.57)$$

---

[9]The Lie algebra $\mathfrak{sp}(4, \mathbb{K})$ also contains the isometry algebra of $\operatorname{Im}(\mathbb{K})$.

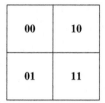 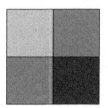

Fig. 11.2   The block structure of a 4×4 antisymmetric matrix in terms of 2×2 blocks. A binary labeling of the blocks is shown on the left; on the right, blocks with similar shading contain equivalent information.

where

$$\epsilon = \begin{pmatrix} 0 & 1 \\ -1 & 0 \end{pmatrix} \tag{11.58}$$

is the Levi-Civita tensor in two dimensions. Using the identity

$$\epsilon \boldsymbol{X} \epsilon = \widetilde{\boldsymbol{X}}^T \tag{11.59}$$

we see that $P = P_0\, \boldsymbol{I} \otimes \epsilon$ takes the form

$$P = \begin{pmatrix} p\,\epsilon & *\boldsymbol{X} \\ -(*\boldsymbol{X})^T & q\,\epsilon \end{pmatrix} \tag{11.60}$$

which is antisymmetric, and whose block structure is shown in Figure 11.2. The diagonal blocks, labeled 00 and 11, are antisymmetric, and correspond to p and q, respectively, whereas the off-diagonal blocks, labeled 01 and 10, contain equivalent information, corresponding to $*\boldsymbol{X}$. Note that $*\boldsymbol{X}$ does not use up all of the degrees of freedom available in an off-diagonal block; the set of *all* antisymmetric 4×4 matrices is *not* an irreducible representation of $\mathfrak{sp}(4, \mathbb{K})$.

The action of $\mathfrak{sp}(4, \mathbb{K})$ on P is given by

$$P \longmapsto AP + PA^T \tag{11.61}$$

for $A \in \mathfrak{sp}(4, \mathbb{K})$, that is, for A satisfying (11.53).[10] When working over $\mathbb{K} = \mathbb{R}$ or $\mathbb{C}$, the action (11.61) is just the antisymmetric square

$$v \wedge w \longmapsto Av \wedge w + v \wedge Aw \tag{11.62}$$

of the natural representation $v \longmapsto Av$, with $v \in \mathbb{K}^4$.

---

[10]Thus, (11.61) can be used if desired to determine the signs in (11.52).

11.5.2 Cubies

Before generalizing the above construction to the 3×3 case, we first consider the analog of $*X$. Let $\mathcal{X} \in \mathbf{H}_3(\mathbb{O})$ be an element of the Albert algebra, which we can regard as a vector-valued 1-form with components $\mathcal{X}_a{}^b$, with $a, b \in \{1, 2, 3\}$. The Hodge dual $*\mathcal{X}$ of $\mathcal{X}$ is a vector-valued 2-form with components

$$(*\mathcal{X})_{abc} = \mathcal{X}_a{}^m \epsilon_{mbc} \tag{11.63}$$

where ϵ_{abc} denotes the Levi-Civita tensor in three dimensions, that is, the completely antisymmetric tensor satisfying

$$\epsilon_{123} = 1 \tag{11.64}$$

and where repeated indices are summed over. We refer to $*\mathcal{X}$ as a *cubie*. We also introduce the dual of ϵ_{abc}, the completely antisymmetric tensor ϵ^{abc} satisfying

$$\epsilon_{mns}\epsilon^{mns} = 6 \tag{11.65}$$

and note the further identities

$$\epsilon_{amn}\,\epsilon^{bmn} = 2\,\delta_a{}^b, \tag{11.66}$$

$$\epsilon_{abm}\,\epsilon^{cdm} = \delta_a{}^c\,\delta_b{}^d - \delta_a{}^d\,\delta_b{}^c, \tag{11.67}$$

$$\epsilon_{abc}\,\epsilon^{def} = \delta_a{}^d\,\delta_b{}^e\,\delta_c{}^f + \delta_b{}^d\,\delta_c{}^e\,\delta_a{}^f + \delta_c{}^d\,\delta_a{}^e\,\delta_b{}^f$$
$$- \delta_a{}^d\,\delta_c{}^e\,\delta_b{}^f - \delta_b{}^d\,\delta_a{}^e\,\delta_c{}^f - \delta_c{}^d\,\delta_a{}^e\,\delta_b{}^f. \tag{11.68}$$

In particular, we have

$$(*\mathcal{X})_{amn}\epsilon^{bmn} = 2\mathcal{X}_a{}^b. \tag{11.69}$$

Operations on the Albert algebra can be rewritten in terms of cubies. For instance,

$$\operatorname{tr}\mathcal{X} = \frac{1}{2}\,\mathcal{X}_{abc}\,\epsilon^{abc}, \tag{11.70}$$

$$\left(*(\mathcal{X}\,\mathcal{Y})\right)_{abc} = \frac{1}{2}\,\mathcal{X}_{amn}\,\mathcal{Y}_{pbc}\,\epsilon^{mnp}, \tag{11.71}$$

$$\left(*(\mathcal{X}\circ\mathcal{Y})\right)_{abc} = \frac{1}{4}\left(\mathcal{X}_{amn}\,\mathcal{Y}_{pbc} + \mathcal{Y}_{amn}\,\mathcal{X}_{pbc}\right)\epsilon^{mnp}, \tag{11.72}$$

$$\operatorname{tr}(\mathcal{X}\circ\mathcal{Y}) = \frac{1}{8}\left(\mathcal{X}_{amn}\,\mathcal{Y}_{pbc} + \mathcal{Y}_{amn}\,\mathcal{X}_{pbc}\right)\epsilon^{mnp}\,\epsilon^{bca}$$
$$= \frac{1}{8}\left(\mathcal{X}_{amn}\,\mathcal{Y}_{pbc} + \mathcal{Y}_{pbc}\,\mathcal{X}_{amn}\right)\epsilon^{mnp}\,\epsilon^{bca}, \tag{11.73}$$

$$(\operatorname{tr}\mathcal{X})(\operatorname{tr}\mathcal{Y}) = \frac{1}{2}\,\mathcal{X}_{abc}\,\mathcal{Y}_{def}\,\epsilon^{abc}\,\epsilon^{def}, \tag{11.74}$$

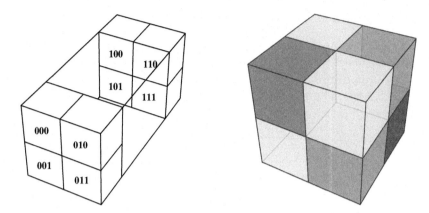

Fig. 11.3   The block structure of a $6 \times 6 \times 6$ antisymmetric tensor in terms of $3 \times 3 \times 3$ "cubies". A binary labeling of the cubies is shown on the pulled-apart cube on the left; on the right, cubies with similar shading contain equivalent information.

from which the components of $*(\mathcal{X} * \mathcal{Y})$ can also be worked out. In the special case where the components of $\mathcal{X}$ and $\mathcal{Y}$ commute, contracting both sides of (11.68) with $\mathcal{X} \otimes \mathcal{Y}$ yields

$$\frac{1}{2} \mathcal{X}_c{}^m \mathcal{Y}_d{}^n \, \epsilon_{amn} \, \epsilon^{bcd} = (\mathcal{X} * \mathcal{Y})_a{}^b \qquad (11.75)$$

or equivalently

$$\left(*(\mathcal{X} * \mathcal{Y}) \right)_{abc} = \frac{1}{2} (\mathcal{X}_b{}^m \mathcal{Y}_c{}^n - \mathcal{X}_c{}^m \mathcal{Y}_b{}^n) \, \epsilon_{amn} \qquad (11.76)$$

providing two remarkably simple expressions for the Freudenthal product, albeit only in a very special case. We will return to this issue below.

11.5.3   *The Symplectic Structure of $\mathfrak{e}_7$*

The representation (11.45) can be written in block form, which we also call Θ, namely[11]

$$\Theta = \begin{pmatrix} \phi - \frac{1}{3} \rho \, \mathcal{I} & \mathcal{A} \\ \mathcal{B} & \phi' + \frac{1}{3} \rho \, \mathcal{I} \end{pmatrix} \qquad (11.77)$$

where $\mathcal{I}$ denotes the 3×3 identity matrix. By analogy with Section 11.5.1, we would like Θ to act on $*\mathcal{X}$, which has three indices, and is antisymmetric in two of them, and hence has the correct symmetries to be an off-diagonal block of a rank three antisymmetric tensor $\mathcal{P}$. The components of $\mathcal{P}$ make

---

[11]The derivations $\mathfrak{g}_2 \subset \mathfrak{e}_6$ require nested matrix transformations of the form (11.77).

up a $6 \times 6 \times 6$ cube, which we divide into $3 \times 3 \times 3$ cubies, as shown in Figure 11.3; compare Figure 11.2. We identity the diagonal cubies, labeled 000 and 111, with $p*\mathcal{I}$ and $q*\mathcal{I}$, respectively, the cubie labeled 011 with $*\mathcal{X}$, the cubie labeled 100 with $*\mathcal{Y}$, and then let antisymmetry do the rest. Explicitly, we have

$$\boldsymbol{P}_{abc} = \begin{cases} p\,\epsilon_{abc} & a \leq 3, b \leq 3, c \leq 3 \\ (*\mathcal{Y})_{\widehat{a}bc} & a \geq 4, b \leq 3, c \leq 3 \\ (*\mathcal{X})_{a\widehat{b}\widehat{c}} & a \leq 3, b \geq 4, c \geq 4 \\ q\,\epsilon_{\widehat{a}\widehat{b}\widehat{c}} & a \geq 4, b \geq 4, c \geq 4 \end{cases} \tag{11.78}$$

where we have introduced the convention that $\widehat{a} = a - 3$, and where the remaining components are determined by antisymmetry.[12]

In the complex case, we could begin with the natural action of Θ on 6-component complex vectors, and then take the antisymmetric cube, that is, we could consider the action

$$u \wedge v \wedge w \longmapsto \Theta u \wedge v \wedge w + u \wedge \Theta v \wedge w + u \wedge v \wedge \Theta w \tag{11.79}$$

with $u, v, w \in \mathbb{C}^6$, or equivalently

$$\boldsymbol{P}_{abc} \longmapsto \Theta_a{}^m \boldsymbol{P}_{mbc} + \Theta_b{}^m \boldsymbol{P}_{amc} + \Theta_c{}^m \boldsymbol{P}_{abm}. \tag{11.80}$$

Over the octonions, however, the action (11.80) must be modified and reinterpreted in order to yield the Freudenthal action (11.47). We summarize the resulting description below; see [7] for further details.

The basic idea is to reorder (11.80), using instead the action

$$\boldsymbol{P}_{abc} \longmapsto \Theta_a{}^m \boldsymbol{P}_{mbc} + \boldsymbol{P}_{amc} \Theta_b{}^m + \boldsymbol{P}_{abm} \Theta_c{}^m \tag{11.81}$$

which, however, is only antisymmetric in its last two indices. We therefore use (11.81) to define the action only on cubies 000, 011, 100, and 111; the action on the remaining four cubies is uniquely determined by requiring that antisymmetry be preserved. This approach succeeds, thanks to the following results [7], which we state without proof.

Lemma 11.1. *The action of $\phi \in \mathfrak{e}_6$ on cubies is given by*

$$(*\mathcal{X})_{abc} \longmapsto \phi_a{}^m (*\mathcal{X})_{mbc} + (*\mathcal{X})_{amc} \phi_b'{}^m + (*\mathcal{X})_{abm} \phi_c'{}^m. \tag{11.82}$$

Lemma 11.2. *The action of the dilation $\Theta = (0, \rho, 0, 0) \in \mathfrak{e}_7$ on $\boldsymbol{P}$ is given by (11.80).*

---

[12]Note that $\boldsymbol{P}$ is a *cube*, and has components $\boldsymbol{P}_{abc}$ with $a, b, c \in \{1, 2, 3, 4, 5, 6\}$, whereas ϵ_{abc}, $*\mathcal{X}_{abc}$, and $*\mathcal{Y}_{abc}$ are the components of *cubies*, which are subblocks of $\boldsymbol{P}$, with $a, b, c \in \{1, 2, 3\}$.

Lemma 11.3. *If the elements of $\mathcal{A}, \mathcal{B} \in \mathbf{H}_3(\mathbb{O})$ commute with those of $\mathcal{P}$, then the action of the translations $\Theta = (0, 0, \mathcal{A}, 0)$ and $\Theta = (0, 0, 0, \mathcal{B})$ on $\mathcal{P}$ is given by (11.80).*

Over $\mathbb{R}$ or $\mathbb{C}$, we're done; Lemmas 11.1, 11.2, and 11.3 together suffice to show that the action (11.80) is the same as the Freudenthal action (11.47)–(11.50). Unfortunately, the action (11.80) fails to satisfy the Jacobi identity over $\mathbb{H}$ or $\mathbb{O}$. However, we can still use Lemmas 11.1, 11.2, and 11.3 to reproduce the Freudenthal action in those cases, as follows.

Lemma 11.4. *The action of $\Theta = (\phi, 0, 0, 0) \in \mathfrak{e}_7$ on $\mathcal{P}$ is determined by (11.81) when acting on elements of the form (11.78), which extends to all of $\mathfrak{e}_7$ by antisymmetry.*

Putting these pieces together, we obtain the final result.

Theorem 11.1. *The Lie algebra $\mathfrak{e}_7$ acts symplectically on cubes, that is, $\mathfrak{e}_6 \subset \mathfrak{e}_7$ acts on cubes via (11.81), as do real translations and the dilation, and all other $\mathfrak{e}_7$ transformations can then be constructed from these transformations using linear combinations and commutators.*

Proof. Lemmas 11.2 and 11.3 are unchanged by the use of (11.81) rather than (11.80), since the components of Θ commute with those of $\mathcal{P}$ in both cases, and Lemma 11.4 verifies that $\mathfrak{e}_6$ acts via (11.81), as claimed. It only remains to show that the remaining generators of $\mathfrak{e}_7$ can be obtained from these elements via commutators.

Using (11.47)–(11.50), it is straightforward to compute the commutator of two $\mathfrak{e}_7$ transformations of the form (11.45). Letting $\phi = \mathcal{Q} \in \mathfrak{e}_6$ be a boost, so that $\mathcal{Q}^\dagger = \mathcal{Q}$ and $\text{tr}(\mathcal{Q}) = 0$, and using the identity

$$-(\mathcal{A} \circ \mathcal{B}) * \mathcal{X} = \left(\mathcal{B} - \text{tr}(\mathcal{B})\mathcal{I}\right) \circ (\mathcal{A} * \mathcal{X}) + \mathcal{A} * (\mathcal{B} \circ \mathcal{X}) \qquad (11.83)$$

for any $\mathcal{A}, \mathcal{B}, \mathcal{X} \in \mathbf{H}_3(\mathbb{O})$, we obtain

$$\left[(0, 0, \mathcal{A}, 0), (\mathcal{Q}, 0, 0, 0)\right] = (0, 0, \mathcal{A} \circ \mathcal{Q}, 0). \qquad (11.84)$$

We can therefore obtain the null translation $(0, 0, \mathcal{Q}, 0)$ for any *tracefree* Albert algebra element $\mathcal{Q}$ as the commutator of $(0, 0, \mathcal{I}, 0)$ and $(\mathcal{Q}, 0, 0, 0)$; a similar argument can be used to construct the null translation $(0, 0, 0, \mathcal{Q})$. $\qquad\square$

Thus, *all* generators of $\mathfrak{e}_7$ can be implemented either as a symplectic transformation on cubes via (11.81), or as the commutator of two such transformations.

11.5.4 Further Properties

We have showed that the algebraic description of the minimal representation of e_7 introduced by Freudenthal corresponds geometrically to a symplectic structure. Along the way, we have emphasized both the similarities and differences between $\mathfrak{e}_7$ and $\mathfrak{so}(10,2)$. Both of these algebras are *conformal*; their elements divide naturally into generalized rotations ($\mathfrak{e}_6$ or $\mathfrak{so}(9,1)$, respectively), translations, and a dilation. Both act naturally on a representation built out of vectors (3×3 or 2×2 Hermitian octonionic matrices, respectively), together with two additional real degrees of freedom (p and q).[13] In the 2×2 case, the representation (11.51) contains just one vector; in the 3×3 case (11.46), there are two. This at first puzzling difference is fully explained by expressing both representations as antisymmetric tensors, as in (11.60) and (11.78), respectively, and as shown geometrically in Figures 11.2 and 11.3.

In the complex case, we have shown that the symplectic action (11.80) exactly reproduces the Freudenthal action (11.47)–(11.50). The analogy goes even further. In $2n$ dimensions, there is a natural map taking two n-forms to a tensor of rank two. When acting on $\mathcal{P}$ (so $n = 3$), this map takes the form

$$\mathcal{P} \longmapsto \mathcal{P}_{acd}\mathcal{P}_{efg}\,\epsilon^{cdefgb} \tag{11.85}$$

where ϵ now denotes the volume element in six dimensions, that is, the completely antisymmetric tensor with $\epsilon^{123456} = 1$. Freudenthal also gives us a "squaring" operation on $\mathfrak{e}_7$, namely the "super-Freudenthal" product $*$ taking elements $\mathcal{P}$ of the minimal representation of $\mathfrak{e}_7$ to elements of $\mathfrak{e}_7$, given by[14]

$$\mathcal{P} * \mathcal{P} = (\phi, \rho, \mathcal{A}, \mathcal{B}) \tag{11.86}$$

---

[13]The relations (11.63) and (11.69) can be used to identify *any* 3×3 matrix over $\mathbb{K}$ with a "cubie-like" piece of a "cube-like" 3-form. However, the set of all such 3-forms does not carry an irreducible representation of e_7; in fact, it does not carry a representation of e_7 at all, as the algebra fails to close. Similar statements hold in the 2×2 case.

[14]We use $*$ to denote this "super-Freudenthal" product because of its analogy to the Freudenthal product $*$, with which there should be no confusion. Neither of these products is the same as the Hodge dual map, also denoted $*$.

where

$$\phi = \langle \mathcal{X}, \mathcal{Y} \rangle, \tag{11.87}$$

$$\rho = -\frac{1}{4}\mathrm{tr}\left(\mathcal{X} \circ \mathcal{Y} - pq\,\mathcal{I}\right), \tag{11.88}$$

$$\mathcal{A} = -\frac{1}{2}\left(\mathcal{Y} * \mathcal{Y} - p\,\mathcal{X}\right), \tag{11.89}$$

$$\mathcal{B} = \frac{1}{2}\left(\mathcal{X} * \mathcal{X} - q\,\mathcal{Y}\right), \tag{11.90}$$

and where

$$\langle X, Y \rangle Z = Y \circ (X \circ Z) - X \circ (Y \circ Z)$$
$$- (X \circ Y) \circ Z + \frac{1}{3}\mathrm{tr}\,(X \circ Y) Z. \tag{11.91}$$

It is not hard to verify that, in the complex case, (11.85) is (a multiple of) $\mathcal{P} * \mathcal{P}$, as given by (11.86)–(11.90).

Furthermore, $\mathfrak{e}_7$ preserves the quartic invariant

$$J = \mathrm{tr}\left((\mathcal{X} * \mathcal{X}) \circ (\mathcal{Y} * \mathcal{Y})\right) - p\det\mathcal{X} - q\det\mathcal{Y}$$
$$- \frac{1}{4}\left(\mathrm{tr}\,(\mathcal{X} \circ \mathcal{Y}) - pq\right)^2 \tag{11.92}$$

which can be constructed using $\mathcal{P} * \mathcal{P}$. Not surprisingly, the quartic invariant (11.92) can be expressed in the complex case as

$$J \sim \mathcal{P}_{gab}\mathcal{P}_{cde}\mathcal{P}_{fhi}\mathcal{P}_{jkl}\,\epsilon^{abcdef}\,\epsilon^{ghijkl} \tag{11.93}$$

up to an overall factor.

Neither the form of the action (11.80), nor the expressions (11.85) and (11.93), hold over $\mathbb{H}$ or $\mathbb{O}$. This failure should not be a surprise, as trilinear tensor products are not well defined over $\mathbb{H}$, let alone $\mathbb{O}$. Nonetheless, Theorem 11.1 does tell us how to extend (11.80) to the octonions. Although it is also possible to write down versions of (11.85) and (11.93) that hold over the octonions, by using case-dependent algorithms to determine the order of multiplication, it is not clear that such expressions have any advantage over the original expressions (11.86)–(11.90) and (11.92) given by Freudenthal.

Despite these drawbacks, it is clear from our construction that $\mathfrak{e}_{7(-25)}$ should be regarded as a natural generalization of the traditional notion of a symplectic Lie algebra, and fully deserves the name $\mathfrak{sp}(6, \mathbb{O})$. Furthermore, given our construction of E_6 in terms of nested matrices acting on the Albert algebra, it is not difficult to extend the treatment given here from the Lie algebra $\mathfrak{e}_7$ to the Lie group E_7, thus showing how E_7 acts on cubies, and justifying the identification $E_7 \cong \mathrm{Sp}(6, \mathbb{O})$.

11.6   The Geometry of E_8

It would be nice to extend our discussion to the last remaining exceptional Lie group, namely E_8. However, unlike the other exceptional Lie groups, E_8 does not have any "small" representations; the minimal representation of E_8 is the adjoint representation, with 248 elements. The minimal representation of G_2 is $\mathbb{O}$ (7-dimensional), F_4 acts on the trace-free part of the Albert algebra (26-dimensional), E_6 acts the Albert algebra (27-dimensional), and E_7 acts on $\{\boldsymbol{\mathcal{P}} = (\mathcal{X}, \mathcal{Y}, p, q)\}$, that is on two copies of the Albert algebra, together with two real numbers (56-dimensional). But E_8 acts on nothing smaller than itself. (More precisely, E_8 acts on its Lie algebra, $\mathfrak{e}_8$.)

The geometry of E_8 is therefore beyond the scope of this book.

PART III
Applications

Chapter 12

Division Algebras in Mathematics

12.1 The Hopf Bundles

Consider a complex column vector

$$v = \begin{pmatrix} b \\ c \end{pmatrix} \tag{12.1}$$

where $b, c \in \mathbb{C}$. What can we do with v? Well, we can take its Hermitian conjugate, defined in Section 7.1, namely

$$v^\dagger = \begin{pmatrix} \bar{b} & \bar{c} \end{pmatrix}. \tag{12.2}$$

We can now multiply v and $v^\dagger$ in two quite different ways. We can produce a number

$$v^\dagger v = |b|^2 + |c|^2 \in \mathbb{R} \tag{12.3}$$

or, if we multiply in the opposite order, a matrix

$$vv^\dagger = \begin{pmatrix} |b|^2 & b\bar{c} \\ c\bar{b} & |c|^2 \end{pmatrix}. \tag{12.4}$$

One way to visualize v is to think of it as a point in $\mathbb{C}^2 = \mathbb{R}^4$, in which case (12.3) is just the usual (squared) vector norm. Suppose we normalize v by requiring that $v^\dagger v = 1$. Then v lies on the unit sphere in $\mathbb{R}^4$, which is 3-dimensional, and hence denoted by $\mathbb{S}^3$.

What about $vv^\dagger$? This matrix is Hermitian! As discussed in Section 9.3, we can therefore identify $vv^\dagger$ with a vector in spacetime, via

$$t + z = |b|^2, \tag{12.5}$$

$$t - z = |c|^2, \tag{12.6}$$

$$x + iy = c\bar{b}. \tag{12.7}$$

We can invert these relationships, obtaining[1]

$$t = \frac{1}{2}\left(|b|^2 + |c|^2\right),$$ (12.8)

$$z = \frac{1}{2}\left(|b|^2 - |c|^2\right),$$ (12.9)

$$x = \frac{1}{2}\left(c\bar{b} + b\bar{c}\right),$$ (12.10)

$$iy = \frac{1}{2}\left(c\bar{b} - b\bar{c}\right).$$ (12.11)

The normalization (12.3) of v tells us that t is constant, which also follows immediately from

$$2t = \operatorname{tr}\left(vv^\dagger\right) = v^\dagger v = 1.$$ (12.12)

So the trace of $vv^\dagger$ is special; what about its determinant? We have

$$\det(vv^\dagger) = |b|^2|c|^2 - (b\bar{c})(c\bar{b}) = |b|^2|c|^2 - |b\bar{c}|^2 = |b|^2|c|^2 - |bc|^2 = 0.$$ (12.13)

Thus, $vv^\dagger$ represents a null vector! Putting this all together, we have

$$x^2 + y^2 + z^2 = t^2 = \text{constant}$$ (12.14)

so that we can think of $vv^\dagger$ as a point on a particular sphere, which we denote by $\mathbb{S}^2$.

We have constructed a famous map between $\mathbb{S}^3$ and $\mathbb{S}^2$ known as the *Hopf map*, which for us is simply given by

$$\mathbb{S}^3 \longrightarrow \mathbb{S}^2,$$

$$v \longmapsto vv^\dagger.$$ (12.15)

This map is not one-to-one. After all, it takes three angles to specify a point in $\mathbb{S}^3$, but only two to specify a point in $\mathbb{S}^2$.

What information have we lost in "squaring" v to get $vv^\dagger$? If we multiply v by a phase, that is, a complex number of the form $e^{i\theta}$, then not only is the norm (12.3) unchanged, but also the square! Explicitly, if

$$w = ve^{i\theta}$$ (12.16)

then

$$ww^\dagger = vv^\dagger.$$ (12.17)

This phase freedom is associated with the circle $\mathbb{S}^1$ of unit complex numbers—which includes antipodal points as a special case. This copy

---

[1] Some authors multiply v by $\sqrt{2}$ to eliminate the factors of 2 in these expressions.

of $\mathbb{S}^1$ is in fact also a group, denoted $U(1)$, since the product of any two unit complex numbers is another such number.

Not surprisingly, the above construction works over any of the division algebras. Note the crucial use of the division algebra property in (12.13) to conclude that $vv^\dagger$ is a null vector! Thus, allowing the components of v in turn to be in $\mathbb{R}$, $\mathbb{C}$, $\mathbb{H}$, and $\mathbb{O}$ results in $vv^\dagger$ being a null vector in 3, 4, 6, or 10 spacetime dimensions, respectively. Imposing the normalization condition (12.3) allows $vv^\dagger$ to be thought of as a point on the spheres $\mathbb{S}^1$, $\mathbb{S}^2$, $\mathbb{S}^4$, and $\mathbb{S}^8$, respectively, while v can be interpreted as a point on the spheres $\mathbb{S}^1$, $\mathbb{S}^3$, $\mathbb{S}^7$, and $\mathbb{S}^{15}$, respectively.

There are a couple of details to check. Over $\mathbb{H}$, it is essential that the phase freedom in (12.16) be written on the right. And of course the phase is given by an arbitrary unit quaternion, corresponding to $\mathbb{S}^3$. Over $\mathbb{O}$, we must be even more careful in identifying the phase freedom; there appear to be associativity issues in the transition from (12.16) to (12.17). However, there is a way around this: $vv^\dagger$ is really complex, since it contains only one octonionic direction. It therefore admits a complex "square root" u. If we apply the phase freedom to u, rather than v, the computation becomes quaternionic; the associativity problems disappear. Explicitly, choosing[2]

$$u = \begin{pmatrix} \frac{b\bar{c}}{|c|} \\ |c| \end{pmatrix} \qquad (12.18)$$

we have

$$uu^\dagger = vv^\dagger \qquad (12.19)$$

and the most general column vector w satisfying (12.17) is

$$w = u\xi \qquad (12.20)$$

where the phase ξ is any unit octonion, which can be thought of as an element of $\mathbb{S}^7$. We can recover v by putting $\xi = \frac{c}{|c|}$.

Putting this all together, we obtain the four Hopf maps

$$\mathbb{S}^{15} \longrightarrow \mathbb{S}^8,$$
$$\mathbb{S}^7 \longrightarrow \mathbb{S}^4,$$
$$\mathbb{S}^3 \longrightarrow \mathbb{S}^2,$$
$$\mathbb{S}^1 \longrightarrow \mathbb{S}^1. \qquad (12.21)$$

It is remarkable that there are precisely four of these maps. It is even more remarkable that they are related to the four division algebras.

---

[2]If $c = 0$, we could choose $u = \begin{pmatrix} |b| \\ 0 \end{pmatrix}$.

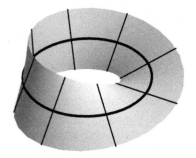

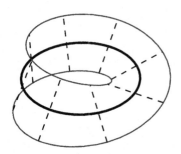

Fig. 12.1   A Möbius strip.   Fig. 12.2   The smallest Hopf bundle.

What do the Hopf maps look like? Consider first a Möbius strip, as shown in Figure 12.1. The Möbius strip is an example of a *fiber bundle*, in which there is a *base manifold*, in this case the circle in the center of the Möbius strip, and a *fiber*, in this case a line segment, otherwise known as the 1-dimensional disk, D^1. The Möbius strip is therefore a fiber bundle over $\mathbb{S}^1$, with fiber D^1. This bundle is nontrivial; it is not simply $\mathbb{S}^1 \times D^1$, which would be a cylinder.

If we now change the fibers by removing all but the endpoints of each line segment, we are left with the fiber bundle shown in Figure 12.2. The base space is still $\mathbb{S}^1$, but the fibers now consist of two points, which is the 0-sphere, S^0, consisting of all points in the line that lie a unit distance from the origin. But these points also make up a circle! Thus, this fiber bundle is itself $\mathbb{S}^1$, over $\mathbb{S}^1$, with fiber S^0—and represents the smallest Hopf map, as given above.

What about the next case, of $\mathbb{S}^3$ over $\mathbb{S}^2$, which is the original Hopf map? The Hopf map is also referred to as the *Hopf bundle*, because it describes $\mathbb{S}^3$ as a fiber bundle with base space $\mathbb{S}^2$ and fiber $\mathbb{S}^1$. Roughly speaking, there is a copy of $\mathbb{S}^1$, given by the phase freedom, at each point of $\mathbb{S}^2$. For the Möbius strip, we obtained the fiber S^0 as the boundary of D^1; we apply the same idea here. Since $\mathbb{S}^3$ is the boundary of D^4, the disk ("ball") in four dimensions, we have

$$\mathbb{S}^3 = \partial D^4 = \partial(D^2 \times D^2)$$
$$= (\partial D^2 \times D^2) \cup (D^2 \times \partial D^2) = (\mathbb{S}^1 \times D^2) \cup (D^2 \times \mathbb{S}^1)$$
$$\cong \mathbb{R}^3 \cup \{\infty\}. \tag{12.22}$$

We can represent $\mathbb{S}^3$ as the points in $\mathbb{R}^3$, together with a point at infinity, as shown in Figure 12.3. There are two disks (D^2) shown in this figure, one inside the torus, the other in the xy plane, in the middle of the torus, but

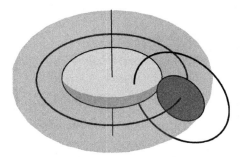

Fig. 12.3   The Hopf bundle.

outside it. Each disk represents a hemisphere of the basespace, with the circles that intersect the disks representing the fibers.[3]

The geometric representations of the first two Hopf bundles shown in Figures 12.2 and 12.3 are purely *topological*, as they fail to show the normalization used in our algebraic construction above.

12.2   The Octonionic Projective Line

There is another way to view $vv^\dagger$, namely as an element in projective space. Consider a pair of real numbers (b, c), and identify points on the same line through the origin. This identification can be thought of as introducing an equivalence relation of the form

$$(b, c) \sim (b\chi, c\chi) \tag{12.23}$$

where $0 \neq \chi \in \mathbb{R}$. The resulting space can be identified with the (unit) circle of all possible directions in $\mathbb{R}^2$, with antipodal points identified. This space is the real *projective space* $\mathbb{RP}^1$, also called the *real projective line*. But this space can also be identified with the squares of *normalized* column vectors v, that is,

$$\mathbb{RP}^1 = \{vv^\dagger : v \in \mathbb{R}^2, v^\dagger v = 1\} \tag{12.24}$$

where we write $\dagger$ instead of T for transpose, anticipating a generalization to the other division algebras.

The normalization condition can be written in terms of the trace, since

$$\operatorname{tr}(vv^\dagger) = v^\dagger v. \tag{12.25}$$

---

[3]It is not easy to see in Figure 12.3 how the two hemispheres are glued together.

There is yet another way to write this condition, since

$$(vv^\dagger)(vv^\dagger) = v(v^\dagger v)v^\dagger = \left(\operatorname{tr}(vv^\dagger)\right)(vv^\dagger). \tag{12.26}$$

Putting the pieces together, we obtain a matrix description of $\mathbb{RP}^1$ in terms of 2×2 real Hermitian matrices $(\mathbf{H}_2(\mathbb{R}))$, namely[4]

$$\mathbb{RP}^1 = \{\boldsymbol{X} \in \mathbf{H}_2(\mathbb{R}) : \boldsymbol{X}^2 = \boldsymbol{X}, \operatorname{tr}\boldsymbol{X} = 1\}. \tag{12.27}$$

Not surprisingly, all of this works over the other division algebras as well; (12.26) holds even over $\mathbb{O}$ since v has only two components, so that the computation takes place in a quaternionic subalgebra. Thus, (12.27) can be used to *define* the projective spaces $\mathbb{RP}^1$, $\mathbb{CP}^1$, $\mathbb{HP}^1$, and $\mathbb{OP}^1$, which are again known as *projective lines*.[5]

However, the traditional definition, using (12.23) rather than (12.27), requires modification over the octonions, along the lines of the discussion in Section 12.1, One possible choice would be

$$\mathbb{OP}^1 = \{(b, c) \in \mathbb{O}^2 : (b, c) \sim \left((bc^{-1})\chi, \chi\right), 0 \neq \chi \in \mathbb{O}\} \tag{12.28}$$

with $c = 0$ handled as a special case.

12.3   Spinors

In Section 9.3, we argued that spacetime vectors are better represented as matrices than as column vectors. What, then, do column vectors represent?

In Section 12.1, we viewed two-component column vectors as vectors in $\mathbb{R}^n$ for appropriate n; imposing a normalization condition led us to regard these vectors as points in $\mathbb{S}^{n-1}$. But this description is really nothing more than counting the (real) degrees of freedom.

A better indication of what these column vectors represent is given by the fact that they "square" to null vectors, in the sense of (12.15). These are *spinors*, more precisely *Weyl* or *Penrose spinors* [14].

How do we "rotate" a spinor? How do we rotate the square root of a vector? Comparing the Lorentz transformation (9.44) with the relationship (12.17) between spinors and (null) vectors yields the answer: A Lorentz transformation $\boldsymbol{M}$ acts on spinors via

$$v \longmapsto \boldsymbol{M}v. \tag{12.29}$$

---

[4]For 2×2 matrices $\boldsymbol{X}$, the condition $\operatorname{tr}\boldsymbol{X} = 1$ is needed to rule out the identity matrix, and ensures that $\det\boldsymbol{X} = 0$, which also forces one of the eigenvalues of $\boldsymbol{X}$ to be 0, which in turn forces $\boldsymbol{X} = vv^\dagger$ for some v.

[5]A similar definition was given by Harvey (page 123 of [13]).

How does the corresponding null vector transform? Equation (9.44) tells us that

$$vv^\dagger \longmapsto M(vv^\dagger)M^\dagger. \tag{12.30}$$

Consistency would require that this vector be the square of Mv, so that

$$M(vv^\dagger)M^\dagger = (Mv)(v^\dagger M^\dagger) = (Mv)(Mv)^\dagger. \tag{12.31}$$

We refer to this condition as *compatibility*. Over $\mathbb{H}$, it is automatically satisfied due to associativity, but over $\mathbb{O}$ it is a nontrivial condition. However, all of the generators given in Section 9.3 do satisfy this condition. *We will always assume that Lorentz transformations are described in terms of a sequence of compatible generators.*

Recall that a generator such as R_z, given in (9.50) in terms of a parameter θ, describes a rotation through an angle 2θ. If $\theta = \pi$, corresponding to a rotation through a full circle, then a vector X is unchanged. However, R_z reduces to $-I$, rather than I. In particular, a spinor is *not* left unchanged by this transformation; it requires a rotation by 4π (corresponding to $\theta = 2\pi$) to do that! This makes sense; since the spinor v is the square root of the vector $vv^\dagger$, it "rotates" half as fast.

But what good are spinors? It turns out that spinors represent particles with half-integer spin. But what does this mean? We will return to the physics of spinors in Chapter 14.

12.4 Möbius Transformations[6]

The unit sphere $\mathbb{S}^2 \subset \mathbb{R}^3$ is related to the Riemann sphere (the complex plane with a point at infinity added) via *stereographic projection* from the north pole, which takes the point (x, y, z), with $x^2 + y^2 + z^2 = 1$, to the point

$$w = \frac{x + iy}{1 - z} = \frac{1 + z}{x - iy}. \tag{12.32}$$

Under this transformation, the north pole is mapped to the point at infinity.

As discussed in detail by Penrose and Rindler [14], we can regard $\mathbb{S}^2$ as the set of future (or past) null directions, and specifically as the intersection of the future light cone of the origin in 4-dimensional Minkowski space with the hypersurface $t = 1$. Other points on a given null ray are obtained by

---

[6]The material in this section is adapted from [15].

scaling with t, and we can extend stereographic projection to a map on the entire light cone via

$$w = \frac{x + iy}{t - z} = \frac{t + z}{x - iy} \qquad (12.33)$$

with the condition $x^2 + y^2 + z^2 = t^2$.

As discussed in Section 12.2, we can further identify $\mathbb{S}^2$ with the complex projective space $\mathbb{CP}^1$, the space of complex lines in $\mathbb{C}^2$, which is given by

$$\mathbb{CP}^1 = \{[(b, c)] \in \mathbb{C}^2 : (b, c) \sim (\xi b, \xi c) \; \forall \, 0 \neq \xi \in \mathbb{C}\} \qquad (12.34)$$

where the square brackets denote equivalence classes under the equivalence relation $\sim$. Then each $[(b, c)] \in \mathbb{CP}^1$ can be identified with the point w in the complex plane given by

$$w = \frac{b}{c} \qquad (12.35)$$

which is further identified with a point in $\mathbb{S}^2$ via (12.32); $[(b, 0)]$ is to be identified with the north pole, corresponding to $w = \infty$. Stereographic projection (12.32) can be thought of as a special case of (12.35) with b or c real.

The Möbius transformations in the complex plane are the complex mappings of the form[7]

$$w \mapsto \frac{\alpha w + \beta}{\gamma w + \delta} \qquad (12.36)$$

where $\alpha\delta - \beta\gamma \neq 0$. It is usually assumed without loss of generality that the complex numbers α, β, γ, δ satisfy

$$\alpha\delta - \beta\gamma = 1. \qquad (12.37)$$

Möbius transformations are the most general analytic transformations of the *Riemann sphere* to itself. Using (12.35), we can rewrite (12.36) as

$$\frac{b}{c} \mapsto \frac{\alpha b + \beta c}{\gamma b + \delta c} \, . \qquad (12.38)$$

The Möbius transformation (12.38) does not depend on the particular choice of b and c in the equivalence class $[(b, c)]$, which allows us to view the transformation as acting on $\mathbb{CP}^1$.

In Section 12.2, we gave an alternate, matrix description of $\mathbb{CP}^1$ as

$$\mathbb{CP}^1 = \{\boldsymbol{X} \in \mathbb{C}^{2 \times 2} : \boldsymbol{X}^\dagger = \boldsymbol{X}, \boldsymbol{X}^2 = \boldsymbol{X}\} \qquad (12.39)$$

---

[7]An excellent description of these transformations, and their relation to Lorentz transformations, appears in §1.2 and §1.3 of [14].

where $X = vv^\dagger$ and $v = \begin{pmatrix} b \\ c \end{pmatrix}$. From this point of view, v is a spinor, whose square X is a null vector. Using these various identifications, we can rewrite a Möbius transformation (12.36) as a map on spinors

$$v \mapsto Mv \tag{12.40}$$

where

$$M = \begin{pmatrix} \alpha & \beta \\ \gamma & \delta \end{pmatrix} \tag{12.41}$$

and, imposing the condition (12.37), we see that $\det M = 1$. We thus see that Möbius transformations are exactly the same as Lorentz transformations. Note the key role played by associativity, which allows one to multiply numerator and denominator of a Möbius transformation by c, thus permitting a reinterpretation as a matrix equation.

But we know how to implement Lorentz transformations over the octonions. In Section 9.3, we presented generators for the Lorentz group, originally given by Manogue and Schray [5], whose components lie in a single complex subalgebra of $\mathbb{O}$ (which may differ for different generators), and whose determinant is real. As discussed in Section 12.3, Such generators also satisfy a *compatibility* condition, namely

$$(Mv)(Mv)^\dagger = M(vv^\dagger)M^\dagger \tag{12.42}$$

between the spinor (v) and vector $(X = vv^\dagger)$ representations.

Putting this all together, we will invert the usual derivation that Lorentz transformations are the same as Möbius transformations. Rather, we will *define* octonionic Möbius transformations in terms of Lorentz transformations, and then show that these transformations can be rewritten in the form (12.38).

Thus, given an octonion w, define (generators of) Möbius transformations via (12.36), which we rewrite as

$$f_M(w) = (\alpha w + \beta)(\gamma w + \delta)^{-1} \tag{12.43}$$

and where the matrix of coefficients M defined by (12.41) is now not only octonionic, but is further required to be one of Manogue & Schray's compatible generators of the Lorentz group.

We would like to be able to construct more general Möbius transformations by nesting. However, it is not at all obvious that iterating (12.43) leads to a (suitably nested) transformation of the same type. We would really like to be able to use (an octonionic version of) (12.38) to define Möbius transformations, as this would make it apparent that iterating Möbius transformations corresponds directly to nesting Lorentz transformations. As previously noted, this construction requires (12.38) to be independent of the particular choice of b and c. Remarkably, the octonionic generalization of (12.38) does have this property, as we now show.

Suppose that

$$w = bc^{-1} \tag{12.44}$$

where now $b, c \in \mathbb{O}$. Letting

$$v_0 = \begin{pmatrix} w \\ 1 \end{pmatrix} \tag{12.45}$$

we have

$$v = v_0 c \tag{12.46}$$

and

$$vv^\dagger = |c|^2 v_0 v_0^\dagger \tag{12.47}$$

since only two octonionic directions are involved.

We now write

$$V = Mv = \begin{pmatrix} B \\ C \end{pmatrix} = \begin{pmatrix} BC^{-1} \\ 1 \end{pmatrix} C \tag{12.48}$$

leading to

$$VV^\dagger = \begin{pmatrix} |B|^2 & B\overline{C} \\ C\overline{B} & |C|^2 \end{pmatrix} = |C|^2 \begin{pmatrix} \frac{|B|^2}{|C|^2} & BC^{-1} \\ \overline{BC^{-1}} & 1 \end{pmatrix} \tag{12.49}$$

and similar relations for $V_0 = Mv_0$. Compatibility now leads to

$$
\begin{aligned}
VV^\dagger &= (Mv)(Mv)^\dagger \\
&= M(vv^\dagger)M^\dagger = |c|^2 M(v_0 v_0^\dagger)M^\dagger \\
&= |c|^2 (Mv_0)(Mv_0)^\dagger = |c|^2 V_0 V_0^\dagger.
\end{aligned} \tag{12.50}
$$

Comparing the offdiagonal entries of (12.50), we obtain

$$|C|^2 BC^{-1} = |c|^2 |C_0|^2 B_0 C_0^{-1}. \tag{12.51}$$

But direct computation shows that

$$|C|^2 = |\gamma b + \delta c|^2 = |\gamma w + \delta|^2 |c|^2 = |C_0|^2 |c|^2 \tag{12.52}$$

provided

$$[b, c, \gamma] \cdot \delta = 0 \tag{12.53}$$

where $\cdot$ denotes the inner product on $\mathbb{O}$ defined in (4.25); this condition is satisfied for compatible M since γ and δ lie in the same complex subspace of $\mathbb{O}$. Finally, by construction we have

$$f_M(w) = B_0 C_0^{-1} \tag{12.54}$$

and putting this all together results in

$$BC^{-1} = f_M(w) \tag{12.55}$$

or equivalently

$$f_M(w) = (\alpha w + \beta)(\gamma w + \delta)^{-1} = (\alpha b + \beta c)(\gamma b + \delta c)^{-1}. \tag{12.56}$$

This is the desired result, since b and c were arbitrary (satisfying (12.44)).

We have shown that the finite octonionic Lorentz transformations in ten dimensions as given by Manogue & Schray [5] can be used to define octonionic Möbius transformations, thus recovering (and correcting) the earlier results of Dündarer, Gürsey, & Tze [16, 17]. However, our approach differs significantly from theirs, as theirs corresponds to using (12.36), while ours uses (12.38). We have thus shown that octonionic Möbius transformations extend to the octonionic projective space $\mathbb{OP}^1$, defined by (12.2). This description could be a key ingredient when attempting to generalize 4-dimensional twistor theory to ten dimensions. Much recent research in superstrings, supergravity, and M-theory has emphasized the importance of lightlike objects in ten dimensions. An appropriate octonionic generalization of twistor theory to ten dimensions might allow powerful twistor techniques to be applied to these other theories.

A key role in our argument is the use of two fundamental properties of the octonionic Lorentz transformations in [5], namely *nesting* and *compatibility*. Our results here support our view that these are essential features of any computation involving octonions. Otherwise, repeated transformations of the form (12.36) are not equivalent to those of the form (12.38), due to the lack of associativity.

12.5   The Octonionic Projective Plane

In Section 12.2, we considered projective *lines*, which are equivalence classes of points in $\mathbb{K}^2$. Those equivalence classes, in turn, correspond to the (normalized) "squares" $(vv^\dagger)$ of the "points" (v). Here, we extend the discussion to projective *planes*, which are described in $\mathbb{K}^3$.

So what is a projective plane? Starting with $\mathbb{R}$, we would like to define an equivalence relation between points, of the form

$$(b, c, r) \sim (b\chi, c\chi, r\chi) \tag{12.57}$$

where $0 \neq \chi \in \mathbb{R}$. Again, it is more natural to consider squares of normalized columns. If we normalize $v \in \mathbb{R}^3$ by

$$v^\dagger v = 1 \tag{12.58}$$

where we have again written $\dagger$ instead of T for transpose, then

$$(vv^\dagger)(vv^\dagger) = v(v^\dagger v)v^\dagger = vv^\dagger. \tag{12.59}$$

Thus, normalized column vectors v have matrix squares $vv^\dagger$ that are projection operators, that is, which square to themselves. Thus, we define

$$\mathbb{RP}^2 = \{\mathcal{X} \in \mathbf{H}_3(\mathbb{R}) : \mathcal{X}^2 = \mathcal{X}, \operatorname{tr}\mathcal{X} = 1\} \tag{12.60}$$

where again the trace condition guarantees that $\mathcal{X} = vv^\dagger$ for some v.

We expect this construction to go through for the other division algebras, and it does, but there's a catch. Over $\mathbb{C}$ and $\mathbb{H}$, everything goes through as expected, yielding projective planes $\mathbb{CP}^2$ and $\mathbb{HP}^2$. What happens over $\mathbb{O}$? The octonionic projective plane, discovered by Ruth Moufang and known as the *Cayley plane*, is indeed defined by

$$\mathbb{OP}^2 = \{\mathcal{X} \in \mathbf{H}_3(\mathbb{O}) : \mathcal{X}^2 = \mathcal{X}, \operatorname{tr}\mathcal{X} = 1\}. \tag{12.61}$$

However, we used associativity in (12.59)! We must therefore ask, which matrices in $\mathbf{H}_3(\mathbb{O})$ square to themselves? The answer is a bit surprising. Writing

$$\mathcal{X} = \begin{pmatrix} p & \bar{a} & c \\ a & m & \bar{b} \\ \bar{c} & b & n \end{pmatrix} \tag{12.62}$$

and squaring, we see that, for instance, the $(1,3)$ component of $\mathcal{X}^2 = \mathcal{X}$ yields

$$(p + n)c + \bar{a}\,\bar{b} = c. \tag{12.63}$$

Since $p, n \in \mathbb{R}$, a, b, c must associate! Thus, the components of matrices in the Cayley plane must lie in a quaternionic subalgebra of $\mathbb{O}$! The components of different matrices can of course lie in different quaternionic subalgebras; nonetheless, this is a significant restriction on the possible matrices.

If we recall the definition of the Freudenthal product given in Section 11.2, we see that

$$\mathcal{X} * \mathcal{X} = \mathcal{X}^2 - \mathcal{X} \operatorname{tr}(\mathcal{X}) + \frac{1}{2}\left(\operatorname{tr}(\mathcal{X})^2 - \operatorname{tr}(\mathcal{X}^2)\right) \tag{12.64}$$

so if $\mathcal{X}^2 = \mathcal{X}$ and $\operatorname{tr}\mathcal{X} = 1$, we obtain

$$\mathcal{X} * \mathcal{X} = 0. \tag{12.65}$$

Conversely, if the Freudenthal square vanishes, then so does its trace

$$\operatorname{tr}(\mathcal{X} * \mathcal{X}) = \frac{1}{2}\left(\operatorname{tr}(\mathcal{X}^2) - \operatorname{tr}(\mathcal{X})^2\right) \tag{12.66}$$

so that (12.65) reduces to

$$\mathcal{X}^2 = \mathcal{X}\operatorname{tr}(\mathcal{X}). \tag{12.67}$$

Thus, points in the Cayley plane are precisely the normalized solutions of (12.65), that is,

$$\mathbb{OP}^2 = \{\mathcal{X} \in \mathbf{H}_3(\mathbb{O}) : \mathcal{X} * \mathcal{X} = 0, \operatorname{tr}\mathcal{X} = 1\}. \tag{12.68}$$

Furthermore, since the components of $\mathcal{X} \in \mathbf{H}_3(\mathbb{O})$ lie in some quaternionic subalgebra $\mathbb{H} \subset \mathbb{O}$ (which depends on $\mathcal{X}$), simple linear algebra shows that

$$\mathcal{X} = vv^\dagger \tag{12.69}$$

for some $v \in \mathbb{H}^3$. Thus, the elements of $\mathbb{OP}^2$ the *squares* of (quaternionic!) triples in $\mathbb{O}^3$.

Now that we know what the *points* in the projective plane are, we need to find the *lines*. We first briefly return to the real case. Let $v, w \in \mathbb{R}^3$ be (real) vectors in three Euclidean dimensions. In Section 11.2, we introduced the Jordan ($\circ$) and Freudenthal ($*$) products on the Albert algebra of octonionic Hermitian 3×3 matrices. We can construct such matrices from v and w, namely vv^T and ww^T. We than have the identities

$$(vv^T) \circ (ww^T) = (v^T w)(vw^T) + (w^T v)(wv^T)$$
$$= (v^T w)(vw^T + wv^T), \tag{12.70}$$
$$\operatorname{tr}(vv^T \circ ww^T) = (v^T w)^2, \tag{12.71}$$
$$(vv^T) * (ww^T) = (v \times w)(v \times w)^T, \tag{12.72}$$

where $v \times w$ denotes the usual cross product in $\mathbb{R}^3$, and where of course $v^T w = w^T v \in \mathbb{R}$. Equations (12.70)–(12.72) justify regarding the trace of the Jordan product as a generalized dot product, and the Freudenthal product as a generalized cross product.

Over $\mathbb{C}$ and $\mathbb{H}$, (12.71) becomes

$$\operatorname{tr}\left(vv^{\dagger} \circ ww^{\dagger}\right) = |v^{\dagger}w|^2 \tag{12.73}$$

since now $v^{\dagger}w = \overline{w^{\dagger}v}$.

Continuing the analogy to vector analysis, a "vector" "cross" itself must vanish. In fact, if we think of $\mathbb{OP}^2 \otimes \mathbb{R}$ as the space of (non-normalized) "vectors", henceforth called "1-squares", so that

$$\mathbb{OP}^2 \otimes \mathbb{R} = \{\mathcal{X} \in \mathbf{H}_3(\mathbb{O}) : \mathcal{X} * \mathcal{X} = 0\} \tag{12.74}$$

then for $\mathcal{A}, \mathcal{B} \in \mathbb{OP}^2 \otimes \mathbb{R}$ we have

$$\mathcal{A} * \mathcal{B} = 0 \iff \mathcal{B} = \lambda\mathcal{A} \tag{12.75}$$

for some $\lambda \in \mathbb{R}$, so that $\mathcal{A}$ and $\mathcal{B}$ are "parallel".

Similar analogies can be made with the dot product. In $\mathbb{HP}^2$ (and hence also in $\mathbb{RP}^2$ and $\mathbb{CP}^2$),

$$\operatorname{tr}\left(vv^{\dagger} \circ ww^{\dagger}\right) = 0 \implies vv^{\dagger} \circ ww^{\dagger} = 0 \tag{12.76}$$

which follows from (12.73). Remarkably, (12.76) still holds in $\mathbb{OP}^2$ even though (12.73) does not, as can be checked by an explicit but lengthy computation. That is,

$$\operatorname{tr}\left(\mathcal{X} \circ \mathcal{Y}\right) = 0 \implies \mathcal{X} \circ \mathcal{Y} = 0 \tag{12.77}$$

for $\mathcal{X}, \mathcal{Y} \in \mathbb{OP}^2$. An interesting consequence of (12.77) is that

$$\mathcal{X} * \mathcal{X} = 0 = \operatorname{tr} \mathcal{X} \iff \mathcal{X} = 0. \tag{12.78}$$

Using our intuition about points and lines in the real projective plane, our knowledge of vector analysis in $\mathbb{R}^3$, and our notions of generalized dot and cross products, we are now ready to study the properties of the octonionic projective plane.

So what are the lines in the projective plane? Recall that the *points* of $\mathbb{OP}^2$ are given by (12.61). Consider by analogy the matrices

$$\Lambda = \{\mathcal{X} \in \mathbf{H}_3(\mathbb{O}) : \mathcal{X}^2 = \mathcal{X}, \operatorname{tr} \mathcal{X} = 2\} \tag{12.79}$$

and call the elements of Λ *lines*.

Why is this terminology reasonable? In algebraic terms, $\mathcal{P} \in \mathbb{OP}^2$ is a *primitive idempotent* of $\mathbf{H}_3(\mathbb{O})$. Restricting temporarily to $\mathbb{H}$ to avoid associativity issues, $\mathcal{P}$ is a projection operator on $\mathbb{H}^3$ into a 1-dimensional subspace, whereas $\mathcal{L} \in \Lambda$ is a projection operator into a 2-dimensional subspace (and $\mathcal{I}$ "projects" into a 3-dimensional subspace, namely all of $\mathbb{H}^3$). In the projective plane, (equivalence classes of) such 1-dimensional subspaces are "points", and (equivalence classes of) such 2-dimensional subspaces are lines. The definitions (12.61) and (12.79) therefore represent plausible generalizations of these concepts to the nonassociative case.

Using definition (12.79), if $\mathcal{Q} \in \mathbb{OP}^2$ is a point, then

$$\mathcal{L} = \mathcal{I} - \mathcal{Q} \in \Lambda \tag{12.80}$$

is a line, where $\mathcal{I}$ is the 3×3 identity matrix, which is also the unique matrix that squares to itself and has trace 3. There is thus a natural duality relationship between the lines and points of $\mathbb{OP}^2$. Using our intuition from $\mathbb{R}^3$, where there is a unique line (through the origin) perpendicular to each plane (through the origin), and vice versa, we can use the point $\mathcal{I} - \mathcal{L}$, which is "perpendicular" to $\mathcal{L}$, to define the points $\mathcal{P}$ that are *on* the line $\mathcal{L}$ as being those points which are orthogonal to $\mathcal{I} - \mathcal{L}$. What does orthogonality mean for points? Use the trace of the Jordan product, that is, define $\mathcal{P}$ to be on $\mathcal{L}$ if

$$\mathrm{tr}\left(\left(\mathcal{I} - \mathcal{L}\right) \circ \mathcal{P}\right) = 0. \tag{12.81}$$

Since $\mathcal{P}, \mathcal{I} - \mathcal{L} \in \mathbb{OP}^2$, we can use (12.77) to drop the trace; $\mathcal{P}$ is on $\mathcal{L}$ if

$$\left(\mathcal{I} - \mathcal{L}\right) \circ \mathcal{P} = 0 \tag{12.82}$$

or equivalently if

$$\mathcal{L} \circ \mathcal{P} = \mathcal{P}. \tag{12.83}$$

What is the line determined by the points $\mathcal{P}$ and $\mathcal{Q}$? Use the cross product! If $\mathcal{P}, \mathcal{Q} \in \mathbb{OP}^2$, then it does indeed follow that

$$\left(\mathcal{P} * \mathcal{Q}\right) * \left(\mathcal{P} * \mathcal{Q}\right) = 0 \tag{12.84}$$

that is, $\mathcal{P} * \mathcal{Q}$ is a 1-square if $\mathcal{P}$ and $\mathcal{Q}$ are 1-squares (and is nonzero so long as $\mathcal{P}$ and $\mathcal{Q}$ are not "parallel"). However, $\mathcal{P} * \mathcal{Q}$ will not in general be normalized. The point "orthogonal" to $\mathcal{P}$ and $\mathcal{Q}$ is therefore $\mathcal{P}*\mathcal{Q}/\mathrm{tr}\left(\mathcal{P}*\mathcal{Q}\right)$, so the line determined by $\mathcal{P}$ and $\mathcal{Q}$ must be[8]

$$\mathcal{L}_{\mathcal{P}\mathcal{Q}} = \mathcal{I} - \frac{\mathcal{P} * \mathcal{Q}}{\mathrm{tr}\left(\mathcal{P} * \mathcal{Q}\right)}. \tag{12.85}$$

---

[8]Some authors, such as Baez [18], consider lines to be given by $\mathcal{P} * \mathcal{Q}$, rather than by expressions such as (12.85).

$\mathcal{P}$ and $\mathcal{Q}$ are on this line, since

$$(\mathcal{P} * \mathcal{Q}) \circ \mathcal{Q} = 0 \tag{12.86}$$

for any 1-squares $\mathcal{P}$ and $\mathcal{Q}$. More generally, three points $\mathcal{P}$, $\mathcal{Q}$, and $\mathcal{S}$ are collinear if

$$(\mathcal{P} * \mathcal{Q}) \circ \mathcal{S} = 0. \tag{12.87}$$

Recall from Sections 11.2–11.4 that E_6 acts on $\mathbf{H}_3(\mathbb{O})$. We briefly summarize some properties of the induced action of E_6 on $\mathbb{OP}^2$.

- The trace of the triple product is the polarization of the determinant, and hence preserved by E_6. Thus, E_6 is precisely the symmetry group which preserves the notion of collinear points in $\mathbb{OP}^2$.
- In fact, E_6 takes p-squares to p-squares, but the boosts in E_6 do not preserve the normalization condition on $\mathbb{OP}^2$. Thus, E_6 takes points to points, and lines to lines *up to normalization* (which can be corrected by slightly modifying the action).
- The action of E_6 on lines is *not* the usual action of E_6 on 2-squares, but rather the "squared" action induced by

$$\mathcal{P} * \mathcal{Q} \longmapsto (\mathcal{MPM}^\dagger) * (\mathcal{MQM}^\dagger),$$

with subsequent renormalization as needed.
- Since F_4 is the automorphism group of the Jordan product, it is also the automorphism group of the Freudenthal product, so that

$$(\mathcal{MPM}^\dagger) * (\mathcal{MQM}^\dagger) = \mathcal{M}(\mathcal{P} * \mathcal{Q})\mathcal{M}^\dagger \tag{12.88}$$

for $\mathcal{M} \in F_4$.
- The remaining elements ("boosts") in E_6 are generated by complex Hermitian matrices, and for such matrices direct computation shows that

$$(\mathcal{MPM}^\dagger) * (\mathcal{MQM}^\dagger) = (\mathcal{M} * \mathcal{M})(\mathcal{P} * \mathcal{Q})(\mathcal{M} * \mathcal{M})^\dagger. \tag{12.89}$$

This operation yields a "dual" action of boosts on 1-squares, given by $\mathcal{M} * \mathcal{M}$ rather than $\mathcal{M}$—and note that $\mathcal{M} * \mathcal{M}$ is (a multiple of) $\mathcal{M}^{-1}$; these boosts go "the other way".
- Any projective line is a 2-square with repeated eigenvalue 1, and can therefore be written (in several ways) as $\mathcal{L} = \mathcal{P} + \mathcal{Q}$, where $\mathcal{P}$, $\mathcal{Q}$ are projective points satisfying $\mathcal{P} \circ \mathcal{Q} = 0$.
- But if $\mathcal{P} \circ \mathcal{Q} = 0$, the Freudenthal product simplifies to

$$2\mathcal{P} * \mathcal{Q} = \mathcal{I} - \mathcal{P} - \mathcal{Q}. \tag{12.90}$$

- Thus, a projective line can be written as $\mathcal{L} = \mathcal{I} - 2\mathcal{P} * \mathcal{Q}$, where $\mathcal{P}$ and $\mathcal{Q}$ are ("orthogonal") points on the line.
- If $\mathcal{L} \in \Lambda$ is a line in $\mathbb{OP}^2$, then the action of E_6 is given by

$$
\mathcal{L} \longmapsto \mathcal{L}' = \begin{cases} \mathcal{I} - \mathcal{M}(\mathcal{I} - \mathcal{L})\mathcal{M}^\dagger & \text{(rotations)} \\ \mathcal{I} - \frac{1}{N}(\mathcal{M} * \mathcal{M})(\mathcal{I} - \mathcal{L})(\mathcal{M} * \mathcal{M})^\dagger & \text{(boosts)} \end{cases}
$$

$$(12.91)$$

where N is a normalization constant.

- Untangling these definitions, the condition that a point $\mathcal{A}$ is on $\mathcal{L}$ is preserved by boosts, since

$$
(\mathcal{M} * \mathcal{M})(\mathcal{I} - \mathcal{L})(\mathcal{M} * \mathcal{M})^\dagger = 2(\mathcal{M} * \mathcal{M})(\mathcal{P} * \mathcal{Q})(\mathcal{M} * \mathcal{M})^\dagger
$$
$$
= 2(\mathcal{M}\mathcal{P}\mathcal{M}^\dagger) * (\mathcal{M}\mathcal{Q}\mathcal{M}^\dagger) \quad (12.92)
$$

and therefore

$$
\mathcal{L}' \circ \mathcal{A}' = \mathcal{A}' \Longleftrightarrow 0 = (\mathcal{M} * \mathcal{M})(\mathcal{I} - \mathcal{L})(\mathcal{M} * \mathcal{M})^\dagger \circ \mathcal{M}\mathcal{A}\mathcal{M}^\dagger
$$
$$
= 2\big((\mathcal{M}\mathcal{P}\mathcal{M}^\dagger) * (\mathcal{M}\mathcal{Q}\mathcal{M}^\dagger)\big) \circ \mathcal{M}\mathcal{A}\mathcal{M}^\dagger
$$
$$
= 2(\mathcal{P} * \mathcal{Q}) \circ \mathcal{A}
$$
$$
\Longleftrightarrow \mathcal{L} \circ \mathcal{A} = \mathcal{A}. \quad (12.93)
$$

- E_6 is therefore the symmetry group which takes points to points and lines to lines in $\mathbb{OP}^2$, while preserving the incidence relation.

12.6 Quaternionic Integers

What are integers? Over $\mathbb{R}$, the answer is easy, namely the infinite set

$$
\mathbb{Z} = \{\ldots, -2, -1, 0, 1, 2, \ldots\}. \quad (12.94)
$$

It is straightforward to extend this definition to the complex numbers, resulting in the *Gaussian integers*

$$
\mathbb{Z}[\ell] = \mathbb{Z} \oplus \mathbb{Z}\ell = \{m + n\ell : m, n \in \mathbb{Z}\}. \quad (12.95)
$$

where we continue to use ℓ rather than i for the complex unit. The Gaussian integers form a *lattice* in two dimensions. The *units* of $\mathbb{Z}[\ell]$ are the elements with norm one, namely the set $\{\pm 1, \pm \ell\}$.

What happens over the other division algebras?

We can of course simply extend the construction of the Gaussian integers. Over $\mathbb{H}$, we obtain the *Lipschitz integers*

$$
\mathbb{Z}[i, j, k] = \mathbb{Z} \oplus \mathbb{Z}i \oplus \mathbb{Z}j \oplus \mathbb{Z}k \quad (12.96)
$$

with units $\{\pm 1, \pm i, \pm j, \pm k\}$. As with both the ordinary integers and the Gaussian integers, the Lipschitz integers are clearly closed under multiplication, and the norm of a Gaussian integer is an (ordinary) integer. Remarkably, the Gaussian integers are not the only subalgebra of $\mathbb{H}$ with these properties.

Consider

$$q = \frac{1}{2}(1 + i + j + k) \in \mathbb{H} \tag{12.97}$$

which has the surprising property that

$$|q| = \frac{1}{4} + \frac{1}{4} + \frac{1}{4} + \frac{1}{4} = 1 \in \mathbb{Z} \tag{12.98}$$

so that the norm of q is an integer, even though the components of q are not. Such "half-integer" quaternions do not quite close under multiplication, since for instance

$$\left(\frac{1}{2}(1 + i + j + k)\right)\left(\frac{1}{2}(1 - i + j + k)\right) = k \tag{12.99}$$

which has integer coefficients rather than half-integer. But the union of the "half-integer" quaternions with the "integer" quaternions, that is, with the Lipschitz integers, *does* close under multiplication. The elements of this union are called *Hurwitz integral quaternions*, or simply *Hurwitz integers*, and can be generated by $\{q, i, j, k\}$, that is, any Hurwitz integer can be written as a product of these four generators. The units among the Hurwitz integers are $\{\pm 1, \pm i, \pm j, \pm k, \frac{1}{2}(\pm 1 \pm i \pm j \pm k)\}$, where all possible combinations of signs are permitted in the last element. As the subalgebra consisting of all rational quaternions with integer norm, there are contexts in which the Hurwitz integers, rather than the Lipschitz integers, play the role of "quaternionic integers".

Why did we add the qualifier "rational"? There are of course other elements of $\mathbb{H}$ and $\mathbb{C}$ with unit norm. For instance, the cube roots of unity over $\mathbb{C}$, which are $\{1, \frac{1}{2}(-1 \pm \sqrt{3}\ell)\}$, not only have unit norm but form a subalgebra of $\mathbb{C}$, that is, they close under multiplication. We could therefore have considered the *Eisenstein integers* $\{m + n\frac{1}{2}(-1 + \sqrt{3}\ell)\}$, which are also called *Euler integers*. However, the unit elements $\pm \ell$ are *not* Eisenstein integers—hence the restriction to rational coefficients.

A remarkable property of the Hurwitz integers is that they contain *rational* cube (and sixth) roots of unity. Because

$$q^3 = -1 \tag{12.100}$$

the Hurwitz integer

$$-q^2 = \frac{1}{2}(1 - k - j - k) \tag{12.101}$$

is a rational cube root of unity.

Further discussion of integral quaternions can be found in [19].

12.7 Octonionic Integers

What are the octonionic integers? Based on our discussion of the quaternionic case in Section 12.6, we require the components of any such octonion to be either an integer or a half-integer, and the norm of any such octonion to be an integer. How many elements of this form are there?

First of all, we have the 16 elements

$$G = \{\pm 1, \pm i, \pm j, \pm k, \pm k\ell, \pm j\ell, \pm i\ell, \pm \ell\}. \tag{12.102}$$

These 16 elements close under multiplication, and generate an 8-dimensional lattice which we will call the *Gravesian integers*, and which generalize the Lipschitz integers over $\mathbb{H}$.[9] However, as is the case over $\mathbb{H}$, the Gravesian integers are not maximal.

Next, we can consider the 2^8 elements

$$\Omega = \left\{ \frac{1}{2}(\pm 1 \pm i \pm j \pm k \pm k\ell \pm j\ell \pm i\ell \pm \ell) \right\} \tag{12.103}$$

where all combinations of signs are allowed. Remarkably, products of elements in Ω either have integer coefficients, such as

$$\frac{1}{2}(1 + i + j + k + k\ell + j\ell + i\ell + \ell)$$

$$\times \frac{1}{2}(1 + i + j + k + k\ell + j\ell + i\ell - \ell)$$

$$= -1 + i + j + k \tag{12.104}$$

or have eight (nonzero) half-integer coefficients, such as

$$\frac{1}{2}(1 + i + j + k + k\ell + j\ell + i\ell + \ell)$$

$$\times \frac{1}{2}(1 - i - j + k + k\ell + j\ell + i\ell + \ell)$$

$$= \frac{1}{2}(-1 + i - j + k + k\ell + 3j\ell + i\ell - \ell)$$

$$= j\ell + \frac{1}{2}(-1 + i - j + k + k\ell + j\ell + i\ell - \ell). \tag{12.105}$$

---

[9]Our terminology for the various sets of integral octonions is adapted from [19].

Thus, the 8-dimensional lattice generated by $G \cup \Omega$, that is, the set of all integer linear combinations of elements in $G \cup \Omega$, also closes under multiplication; these are the *Kleinian integers*. However, the Kleinian integers are also not maximal.

What happened to the Hurwitz integers? Surely we should include elements such as $q = \frac{1}{2}(1+i+j+k)$ and, once we do so, also the "complement" of such elements, such as $\frac{1}{2}(k\ell + j\ell + i\ell + \ell)$. There's only one problem: The set generated by the $2 \times 7 \times 2^4 = 224$ such elements, together with G and Ω, collectively known as the *Kirmse integers*, doesn't close under multiplication! For example,

$$\frac{1}{2}(1 + i + j + k) \times \frac{1}{2}(1 + k + k\ell + \ell) = \frac{1}{2}(i + k + k\ell + j\ell) \qquad (12.106)$$

which has not yet been included (since the complement of $\{i, k, k\ell, j\ell\}$ is $\{1, j, i\ell, \ell\}$, which does *not* correspond to a quaternionic subalgebra of $\mathbb{O}$). Even if we include such elements, expanding our 224 Kirmse units to a set of $\binom{8}{4} \times 2^4 = 1120$ units, the product still doesn't close. For example,

$$\frac{1}{2}(i + k + k\ell + j\ell) \times \frac{1}{2}(1 + i + j\ell + k\ell)$$

$$= \frac{1}{4}(-3 + i + j + k + k\ell + j\ell + k\ell - \ell) \qquad (12.107)$$

which is clearly not in our pool of candidate integral octonions.

So how do we generalize the Hurwitz integers? Let's start again. Recall that the Hurwitz units over $\mathbb{H}$ are

$$H = \{\pm 1, \pm i, \pm j, \pm k, \frac{1}{2}(\pm 1 \pm i \pm j \pm k)\}. \qquad (12.108)$$

If we now include the Gravesian units, we can generate the complement of H as $H\ell$, and the Kleinian elements Ω as $H \oplus H\ell$. The algebra generated by $H \cup G \cup \Omega$ closes, since all products that don't involve Ω lie entirely in $\mathbb{H}$ or $\mathbb{H}^{\perp}$. We will refer to this algebra as the *double Hurwitzian integers* generated by $\mathbb{H}$. Any of the other six quaternionic triples can be used in place of $\{i, j, k\}$; there are seven different copies of the double Hurwitzian integers in $\mathbb{O}$. However, the double Hurwitzian integers are still not maximal.

As discussed in [19], the trick to obtaining a maximal set of integral octonions is to choose a preferred octonionic unit; we choose ℓ. Looking at the octonionic multiplication table, there are three quaternionic triples containing ℓ, so include the Hurwitz integers corresponding to those three quaternionic subalgebras, but not those corresponding to the other four quaternionic subalgebras. As before, include their complements—which can also be obtained as sums and differences of Hurwitz integers from different quaternionic subalgebras. We're almost there. Computing products, we discover that we have generated a different set of 224 "generalized Kirmse units", namely

$$
\Omega_0 = \Big\{ \frac{1}{2}(\pm 1 \pm i \pm i\ell \pm \ell), \frac{1}{2}(\pm 1 \pm j \pm j\ell \pm \ell), \frac{1}{2}(\pm 1 \pm k \pm k\ell \pm \ell),
$$
$$
\frac{1}{2}(\pm j \pm j\ell \pm k \pm k\ell), \frac{1}{2}(\pm k \pm k\ell \pm i \pm i\ell), \frac{1}{2}(\pm i \pm i\ell \pm j \pm j\ell),
$$
$$
\frac{1}{2}(\pm 1 \pm i\ell \pm j \pm k), \frac{1}{2}(\pm 1 \pm i \pm j\ell \pm k), \frac{1}{2}(\pm 1 \pm i \pm j \pm k\ell),
$$
$$
\frac{1}{2}(\pm i \pm j\ell \pm k\ell \pm \ell), \frac{1}{2}(\pm i\ell \pm j \pm k\ell \pm \ell), \frac{1}{2}(\pm i\ell \pm j\ell \pm k \pm \ell),
$$
$$
\frac{1}{2}(\pm 1 \pm i\ell \pm j\ell \pm k\ell), \frac{1}{2}(\pm i \pm j \pm k \pm \ell) \Big\}. \tag{12.109}
$$

Remarkably, the lattice generated by $G \cup \Omega \cup \Omega_0$ does close under multiplication; we call its elements the "ℓ-integers". There are $224 + 16 = 240$ units in the ℓ-integers, since the elements of Ω are sums of elements of Ω_0 (and are not units anyway). Furthermore, the ℓ-integers are maximal; no further elements can be added without giving up either closure under multiplication or our defining properties for candidate integers (namely integer norm and half-integer or integer coefficients).

This construction can of course be repeated replacing ℓ with any other imaginary octonionic unit; there are thus seven different maximal "orders" of integral octonions.

Further information about integral octonions can be found in [19].

12.8 The Geometry of the $\mathfrak{e}_8$ Lattice

The octonionic integers provide a natural description of the $\mathfrak{e}_8$ lattice, as we now show.

Recall from Section 10.3 that a simple Lie algebra of dimension d and rank r contains a Cartan subalgebra of dimension r, consisting of mutually consisting elements. Over $\mathbb{C}$, the remaining $d - r$ elements can be chosen to be simultaneous eigenvectors of Cartan elements, with real eigenvalues; the resulting r-tuples form the *root diagram* of the Lie algebra, which generates a lattice.

It is intriguing that the root diagram of $\mathfrak{e}_8$, and the corresponding lattice, being 8-dimensional, can be represented using octonions. Since the dimension of $\mathfrak{e}_8$ is 248, and the rank is 8, the root diagram has 240 elements.

One representation of the $\mathfrak{e}_8$ lattice is obtained by first considering the $2^8 = 256$ elements in Ω (see (12.103)). Clearly, half of these elements have an even number of minus signs; call the collection of such elements Ω_+. We next consider $g_1 + g_2$ for pairs of elements $g_1, g_2 \in G$ (see (12.102)) that satisfy $g_1 \neq \pm g_2$; there are $\binom{8}{2} \times 2^2 = 112$ such elements, which we collectively call G_0. Then the root diagram of $\mathfrak{e}_8$ can be given by $\Omega_+ \cup G_0$, which generates the *odd* $\mathfrak{e}_8$ lattice, which we will call L, following [20]. The conjugate lattice $R = \overline{L}$ yields another representation of the $\mathfrak{e}_8$ lattice.

A somewhat different representation of the $\mathfrak{e}_8$ lattice is obtained by considering the union of the Hurwitz integers corresponding to all seven quaternionic triples, as well as their complements, namely the 224 Kirmse integers, together with the 16 elements of G. This construction yields a third copy of the $\mathfrak{e}_8$ root diagram; the corresponding lattice is the *even* $\mathfrak{e}_8$ lattice, denoted B.

None of the lattices L, R, B closes under multiplication. However, each of these lattices is closely related to integral octonions! For instance, choose an octonion (ℓ) to be special, define

$$a = \frac{1}{2}(1 + \ell) \tag{12.110}$$

and denote the ℓ-integers of Section 12.7 by A. Then

$$A = aL = Ra = 2aBa \tag{12.111}$$

so that

$$L = 2aA = 2\overline{a}A, \tag{12.112}$$

$$R = 2Aa = 2A\overline{a}, \tag{12.113}$$

$$B = La = aR = 2aAa, \tag{12.114}$$

where we have used the fact that $a = 2a^2\overline{a}$, with $2a^2 = \ell \in A$. As discussed by Wilson [20], the Moufang identities now imply that

$$BL = L, \tag{12.115}$$

$$RB = R, \tag{12.116}$$

$$LR = 2B. \tag{12.117}$$

Thus, the $\mathfrak{e}_8$ lattice has the remarkable property that it can (also) be represented as an *integral subalgebra* of $\mathbb{O}$, namely A.

Analogous descriptions exist for the other division algebras. The (rescaled) root diagram of $\mathfrak{a}_1 = \mathfrak{su}(2)$ consists of the two real numbers ± 1, which is clearly closed under multiplication. The (rescaled) root diagram of $\mathfrak{a}_2 = \mathfrak{su}(3)$ is a hexagon, which can be represented as the complex sixth roots of unity—which is again closed under multiplication. Finally, the root diagram of $\mathfrak{d}_4 = \mathfrak{so}(8)$ is normally given as the $\binom{4}{2} \times 2^2 = 24$ quaternionic vectors in G_0, but a simple rotation (and renormalization) turns this into a copy of the Hurwitz integers—which are the rational quaternionic sixth roots of unity.

Further information about using octonions to describe the $\mathfrak{e}_8$ lattice can be found in [20].

Chapter 13

Octonionic Eigenvalue Problems

13.1 The Eigenvalue Problem

The eigenvalue problem as usually stated is to find *eigenvectors* $v \neq 0$ and *eigenvalues* λ that satisfy

$$Av = \lambda v \tag{13.1}$$

for a given square matrix A, which we will assume to be complex and Hermitian ($A^\dagger = A$). The basic properties of the eigenvalue problem for such matrices are well-understood:

(1) *The eigenvalues of complex Hermitian matrices are real.*
(2) *Eigenvectors of a complex Hermitian matrix corresponding to different eigenvalues are orthogonal.*
(3) *The eigenvectors of any complex Hermitian matrix (can be chosen to) form an orthonormal basis.*
(4) *Any complex Hermitian matrix admits a decomposition in terms of an orthonormal basis of eigenvectors.*

Property 1 follows by computing

$$\overline{\lambda} v^\dagger v = (Av)^\dagger v = v^\dagger A v = \lambda v^\dagger v \tag{13.2}$$

so that if $v \neq 0$ we have $0 \neq v^\dagger v \in \mathbb{R}$, which forces $\overline{\lambda} = \lambda$. Property 2 follows by considering eigenvectors v_m with eigenvalues λ_m. By Property 1, $\lambda_m \in \mathbb{R}$. Then

$$\lambda_1 v_1^\dagger v_2 = (Av_1)^\dagger v_2 = v_1^\dagger A v_2 = \lambda_2 v_1^\dagger v_2 \tag{13.3}$$

so if $\lambda_1 \neq \lambda_2$ we must have $v_1^\dagger v_2 = 0$, which defines orthogonality.

Property 3 follows from the previous properties using the additional fact that there are exactly the right number of independent eigenvectors to form a basis, which we will not prove.[1] Property 4 is the statement that $\boldsymbol{A}$ can be expanded as

$$\boldsymbol{A} = \sum_{m=1}^{n} \lambda_m v_m v_m^\dagger \qquad (13.4)$$

where $\{v_m\}$ is an orthonormal basis of eigenvectors corresponding to the eigenvalues λ_m, that is, where

$$\boldsymbol{A}v_m = \lambda_m v_m, \qquad (13.5)$$

$$v_m^\dagger v_n = \delta_{mn}, \qquad (13.6)$$

where δ_{mn} is the Kronecker delta. This expansion follows from the other properties by checking that

$$\left(\sum_{m=1}^{n} \lambda_m v_m v_m^\dagger \right) v_k = \lambda_k v_k. \qquad (13.7)$$

Which of these properties, if any, hold over the other division algebras?

Over the quaternions, there are Hermitian matrices which admit non-real eigenvalues. For instance, we have

$$\begin{pmatrix} 0 & -i \\ i & 0 \end{pmatrix} \begin{pmatrix} 1 \\ k \end{pmatrix} = \begin{pmatrix} j \\ i \end{pmatrix} = j \begin{pmatrix} 1 \\ k \end{pmatrix}. \qquad (13.8)$$

What went wrong? The proof of Property 1 uses commutativity to move the eigenvalue λ around; this is no longer valid. Is there a way around this?

A bit of thought reveals that (13.1) is no longer the only eigenvalue equation. The right eigenvalue problem turns out to be, well, the right eigenvalue problem, that is

$$\boldsymbol{A}v = v\lambda. \qquad (13.9)$$

If the quaternionic matrix $\boldsymbol{A}$ is Hermitian, careful computation shows that

$$\overline{\lambda}(v^\dagger v) = (\overline{\lambda} v^\dagger) v = (\boldsymbol{A}v)^\dagger v = (v^\dagger \boldsymbol{A}) v$$
$$= v^\dagger (\boldsymbol{A}v) = v^\dagger (v\lambda) = (v^\dagger v)\lambda \qquad (13.10)$$

which uses associativity, but not commutativity. Since $v^\dagger v \in \mathbb{R}$, we can still conclude that $\lambda \in \mathbb{R}$. Similarly, orthogonality follows from

$$\lambda_1 (v_1^\dagger v_2) = (\lambda_1 v_1^\dagger) v_2 = (\boldsymbol{A}v_1)^\dagger v_2 = (v_1^\dagger \boldsymbol{A}) v_2$$
$$= v_1^\dagger (\boldsymbol{A}v_2) = v_1^\dagger (v_2 \lambda_2) = (v_1^\dagger v_2)\lambda_2 \qquad (13.11)$$

(and the fact that $\lambda_m \in \mathbb{R}$). Properties 1–4 therefore hold over the quaternions, so long as the eigenvalues are written on the right, as in (13.9). What happens over the octonions?

---

[1]The Gram–Schmidt orthogonalization process can be used on any eigenspace corresponding to an eigenvalue with multiplicity greater than one.

The use of associativity in the last two derivations leads one to suspect that something will go wrong. It does; even the right eigenvalues of octonionic Hermitian matrices need not be real. For instance, we have

$$\begin{pmatrix} 0 & -i \\ i & 0 \end{pmatrix} \begin{pmatrix} j \\ \ell \end{pmatrix} = \begin{pmatrix} -i\ell \\ k \end{pmatrix} = \begin{pmatrix} j \\ \ell \end{pmatrix} k\ell. \tag{13.12}$$

Nevertheless, it turns out that there is a sense in which all of Properties 1–4 hold over the octonions, at least for 2×2 and 3×3 octonionic Hermitian matrices [21]. We discuss each of these cases in turn.

13.2 The 2×2 Real Eigenvalue Problem

We consider the eigenvector problem $Av = v\lambda$ over $\mathbb{O}$, and look (only) for solutions with real eigenvalues, that is, where $\lambda \in \mathbb{R}$.

There are no surprises in this case, or rather the only surprise is that we cannot show that λ is real, but must assume this property separately. First of all, there is only one independent octonion in A; the components of A live in a complex subalgebra $\mathbb{C} \subset \mathbb{O}$. We can write

$$A = \begin{pmatrix} p & \bar{a} \\ a & m \end{pmatrix} \tag{13.13}$$

with $a \in \mathbb{C}$, $p, m \in \mathbb{R}$, and we will assume $a \neq 0$. Setting

$$v = \begin{pmatrix} x \\ y \end{pmatrix} \tag{13.14}$$

with $x, y \in \mathbb{O}$ brings the eigenvalue equation to the form

$$px + \bar{a}y = \lambda x, \tag{13.15}$$

$$ax + my = \lambda y, \tag{13.16}$$

either of which suffices to show that a, x, y associate, and therefore lie in some quaternionic subalgebra $\mathbb{H} \subset \mathbb{O}$. So at first sight, the eigenvalue problem for 2×2 octonionic Hermitian matrices with real eigenvalues reduces to the quaternionic case.

Not so fast. Each eigenvector must indeed be quaternionic in the sense above, but different eigenvectors can, together with A, determine *different* quaternionic subalgebras.

Let's try again. Since A is complex, we can start by solving the *complex* eigenvalue problem, which allows us to write

$$A = \lambda_1 v_1 v_1^\dagger + \lambda_2 v_2 v_2^\dagger \tag{13.17}$$

where the λ_m are real, and where the components of both $\boldsymbol{A}$ and the v_m all lie in $\mathbb{C}$. If we now set

$$w_m = v_m \xi_m \tag{13.18}$$

for any *octonions* $\xi_m \in \mathbb{O}$, it's clear that w_m is still an eigenvector of $\boldsymbol{A}$ with eigenvalue λ_m; there are no associativity issues here, since $\boldsymbol{A}$ and (a single) ξ_m determine (at most) a quaternionic subalgebra of $\mathbb{O}$. Furthermore, if we normalize the ξ_m by setting

$$|\xi_m| = 1 \tag{13.19}$$

then we can replace v_m by w_m in (13.17) without changing anything.

We claim that all eigenvectors v of $\boldsymbol{A}$ have the form (13.18). One way to see this is to note that, over $\mathbb{H}$, we can make at least one component of v real by right-multiplying by a suitable (quaternionic) phase, which does not otherwise affect the eigenvalue equation. But the eigenvalue equation then forces the other component to lie in $\mathbb{C}$, not merely in $\mathbb{H}$. Thus, the result of multiplying v by this phase is one of the *complex* eigenvectors of $\boldsymbol{A}$. Reversing this process establishes the claim.

It is instructive to work out these properties using explicit components. From (13.15) and (13.16) we obtain

$$(\lambda - p)(\lambda - m)x = \bar{a}(\lambda - m)y = \bar{a}(ax) = |a|^2 x \tag{13.20}$$

where we have used the fact that $p, m, \lambda \in \mathbb{R}$, so that

$$\left((\lambda - p)(\lambda - m) - |a|^2\right) x = 0. \tag{13.21}$$

Assuming $x \neq 0$ (if not, start over with y), we recover the usual *characteristic equation*

$$\det(\boldsymbol{A} - \lambda \boldsymbol{I}) = 0 \tag{13.22}$$

for the eigenvalues of $\boldsymbol{A}$. The rest is easy; at least one of (13.15) and (13.16) determines y in terms of x. None of these manipulations use either commutativity or associativity, although alternativity is required in the last equality of (13.20).

Putting this all together, the eigenvectors of the 2×2 octonionic Hermitian matrix (13.13) with eigenvalue λ can be given in either of the forms

$$v = \begin{pmatrix} |a|^2 \\ a(\lambda - p) \end{pmatrix} \xi, \qquad v = \begin{pmatrix} \bar{a}(\lambda - m) \\ |a|^2 \end{pmatrix} \xi \tag{13.23}$$

where of course λ must solve (13.22).

There is however one last surprise. Associativity hasn't been a problem yet, since a single eigenvector involves only a and ξ, and hence has components that live in a quaternionic subalgebra of $\mathbb{O}$. So if v_m are the complex eigenvectors of A corresponding to eigenvalues λ_m, with $m = 1, 2$, then of course

$$v_1^\dagger v_2 = 0 \tag{13.24}$$

that is, v_1 is orthogonal to v_2. However, with w_m as in (13.18), we have

$$w_1^\dagger w_2 = (v_1 \xi_1)^\dagger (v_2 \xi_2) = (\bar{\xi}_1 v_1^\dagger)(v_2 \xi_2) \tag{13.25}$$

which is *not* necessarily zero, since we cannot in general move the parentheses. A simple counterexample can be constructed using

$$v_1 = \begin{pmatrix} 1 \\ i \end{pmatrix}, \quad v_2 = \begin{pmatrix} 1 \\ -i \end{pmatrix}, \tag{13.26}$$

and setting $\xi_1 = j$, $\xi_2 = \ell$. So what does orthogonality mean?

The answer is both simple and elegant, and can be motivated by looking again at (13.17), in which we can use either v_m or w_m (assuming that $|\xi_m| = 1$). Why? Because ξ_m cancels out in the square! This suggests that the correct notion of orthogonality is given by

$$v \perp w \iff (vv^\dagger)w = 0 \tag{13.27}$$

and it is easy to see that this notion of orthogonality does indeed hold between any two eigenvectors of A with different eigenvalues. In the associative case, we can move the parentheses in (13.27), but not in general.

Further details can be found in [21].

13.3   The 2 × 2 Non-real Eigenvalue Problem[2]

The general 2×2 octonionic Hermitian matrix can be written as (13.13) with $p, m \in \mathbb{R}$ and $a \in \mathbb{O}$, and satisfies its characteristic equation (13.22), which can be written as

$$A^2 - (\operatorname{tr} A) A + (\det A) I = 0 \tag{13.28}$$

where $\operatorname{tr} A = p + m$ denotes the *trace* of A, and where there is no difficulty in defining the determinant of A as usual via

$$\det A = pm - |a|^2 \tag{13.29}$$

since the components of A lie in a complex subalgebra $\mathbb{C} \subset \mathbb{O}$. If $a = 0$ the eigenvalue problem is trivial, so we assume $a \neq 0$. We also write v in the form (13.14), with $x, y \in \mathbb{O}$.

---

[2]The material in this section is adapted from [22].

13.3.1 Left Eigenvalue Problem

As pointed out in Section 13.1, even quaternionic Hermitian matrices can admit left eigenvalues which are not real, as is shown by the following example:

$$\begin{pmatrix} 1 & -i \\ i & 1 \end{pmatrix} \begin{pmatrix} 1 \\ k \end{pmatrix} = \begin{pmatrix} 1 + j \\ k + i \end{pmatrix} = (1 + j) \begin{pmatrix} 1 \\ k \end{pmatrix}. \tag{13.30}$$

Direct computation allows us to determine which Hermitian matrices A admit left eigenvalues which are not real. Inserting (13.14) into the left eigenvalue equation (13.1) leads to

$$(\lambda - p)x = ay, \qquad (\lambda - m)y = \bar{a}x, \tag{13.31}$$

which in turn leads to

$$\frac{\bar{a}\left((\lambda - p)x \right)}{|a|^2} = y = \frac{(\bar{\lambda} - m)(\bar{a}x)}{|\lambda - m|^2}. \tag{13.32}$$

Assuming without loss of generality that $x \neq 0$ and taking the norm of both sides yields

$$|a|^2 = |\lambda - p||\lambda - m| \tag{13.33}$$

resulting in

$$\frac{\bar{a}\left((\lambda - p)x \right)}{|\lambda - p|} = \frac{(\bar{\lambda} - m)(\bar{a}x)}{|\lambda - m|}. \tag{13.34}$$

This equation splits into two independent parts, the terms (in the numerator) which involve the imaginary part of λ, which is nonzero by assumption,[3] and those which don't. Looking first at the latter leads to

$$p = m \tag{13.35}$$

which in turn reduces (13.34) to

$$\bar{a}(\lambda x) = \bar{\lambda}(\bar{a}x) \tag{13.36}$$

which forces a to be purely imaginary (and orthogonal to λ), but which puts no conditions on x.

Denoting the 2×2 identity matrix by I, setting

$$J(r) = \begin{pmatrix} 0 & -r \\ r & 0 \end{pmatrix} \tag{13.37}$$

---

[3]We assume without loss of generality that $\mathrm{Re}\,(\lambda) = 0$ by replacing A with $A - \mathrm{Re}\,(\lambda)I$.

and noting that r is a pure imaginary unit octonion if and only if $r^2 = -1$, we have:

Lemma 13.1. *The set of 2×2 Hermitian matrices A for which non-real left eigenvalues exist is*

$$A_2 = \{A = p\,I + q\,J(r);\ p, q \in \mathbb{R},\ q \neq 0,\ r^2 = -1\}. \tag{13.38}$$

The set A_2 has some remarkable properties, which will be further discussed below. Without loss of generality, we can take $r = i$, so that A takes the form

$$A = \begin{pmatrix} p & -iq \\ iq & p \end{pmatrix}. \tag{13.39}$$

Let us find the general solution of the left eigenvalue problem for these matrices. Taking A as in (13.39) and v as in (13.14), the left eigenvalue equation becomes

$$\frac{\lambda - p}{q}\, x = -iy, \tag{13.40}$$

$$\frac{\lambda - p}{q}\, y = ix. \tag{13.41}$$

Taking the norm of both sides immediately yields

$$|x|^2 = |y|^2 \tag{13.42}$$

and we can normalize both of these to 1 without loss of generality. We thus obtain

$$\frac{\lambda - p}{q} = -(iy)\overline{x} = (ix)\overline{y}$$

$$= -[i, y, \overline{x}] - i(y\overline{x}) = [i, x, \overline{y}] + i(x\overline{y}). \tag{13.43}$$

But since

$$[z, y, \overline{x}] = -[z, y, x] = [z, x, y] = -[z, x, \overline{y}] \tag{13.44}$$

for any z, the two associators are equal, and we are left with

$$x \cdot y = x\overline{y} + y\overline{x} = 0. \tag{13.45}$$

Thus, x and y correspond to orthonormal vectors in $\mathbb{O}$ thought of as $\mathbb{R}^8$. This argument is fully reversible; any suitably normalized x and y which are orthogonal yield an eigenvector of A. We have therefore shown that *all* matrices in A_2 have the same left eigenvectors:

Lemma 13.2. *The set of left eigenvectors for any matrix $A \in A_2$ is*

$$V_2 = \left\{ \begin{pmatrix} x \\ y \end{pmatrix} : |x|^2 = |y|^2;\ x \cdot y = 0 \right\}. \tag{13.46}$$

Given x and y, the left eigenvalue is given in each case by either (13.40) or (13.41). Furthermore, left multiplication by an arbitrary octonion preserves the set V_2, so that matrices in A_2 have the property that left multiplication of left eigenvectors yields another left eigenvector (albeit with a different eigenvalue).[4] It follows from (13.40) or (13.41) and (13.42) that

$$|\lambda - p| = q. \tag{13.47}$$

Inserting (13.47) into either (13.40) or (13.41), multiplying both sides by i, and using the identities

$$a \cdot (xb) = b \cdot (\overline{x}a), \tag{13.48}$$

$$(ax) \cdot (bx) = |x|^2 \, a \cdot b, \tag{13.49}$$

for any $a, b, x \in \mathbb{O}$ (where $\cdot$ denotes the inner product on $\mathbb{O}$ defined in (4.25)), then shows that (13.45) forces $\lambda \cdot i = 0$, or more generally

$$\lambda \cdot a = 0. \tag{13.50}$$

However, (13.47) and (13.50) are the only restrictions on λ, in the sense that (13.40) or (13.41) can be used to construct eigenvectors having *any* eigenvalue satisfying these two conditions.

13.3.2   *Right Eigenvalue Problem*

As discussed in Section 13.1, the *right* eigenvalues of quaternionic Hermitian matrices must be real, which is a strong argument in favor of putting the eigenvalues on the right. However, as also pointed out Section 13.1, there do exist *octonionic* Hermitian matrices which admit right eigenvalues which are not real, as is shown by the following example:

$$\begin{pmatrix} 1 & -i \\ i & 1 \end{pmatrix} \begin{pmatrix} j \\ \ell \end{pmatrix} = \begin{pmatrix} j - i\ell \\ \ell + k \end{pmatrix} = \begin{pmatrix} j \\ \ell \end{pmatrix} (1 + k\ell). \tag{13.51}$$

Proceeding as we did for left eigenvectors, we can determine which matrices $\mathbf{A}$ admit right eigenvalues which are not real. Inserting (13.14) into the right eigenvalue equation leads to

$$x(\lambda - p) = ay, \quad y(\lambda - m) = \overline{a}x, \tag{13.52}$$

which in turn leads to

$$\frac{\overline{a}\Big(x(\lambda - p)\Big)}{|a|^2} = y = \frac{(\overline{a}x)(\overline{\lambda} - m)}{|\lambda - m|^2}. \tag{13.53}$$

---

[4]Direct computation shows that, other than real matrices, the matrices in A_2 are the only 2×2 Hermitian matrices with this property.

Taking the norm of both sides (and assuming $x \neq 0$) again yields (13.33), resulting in

$$\frac{\overline{a}\Big(x(\lambda - p)\Big)}{|\lambda - p|} = \frac{(\overline{a}x)(\overline{\lambda} - m)}{|\lambda - m|}. \tag{13.54}$$

Just as for the left eigenvector problem, this equation splits into two independent parts, the terms (in the numerator) which involve the imaginary part of λ, which is nonzero by assumption, and those which don't. Looking first at the latter again forces $p = m$, which in turn forces $|y| = |x|$. The remaining condition is now

$$\overline{a}(x\lambda) = (\overline{a}x)\overline{\lambda} \tag{13.55}$$

so that $\overline{a}$, $\mathrm{Im}\,(\lambda)$, and x antiassociate. In particular, this forces both a and x to be pure imaginary, as well as

$$\lambda \cdot a = 0, \tag{13.56}$$
$$\lambda \cdot x = 0 = a \cdot x, \tag{13.57}$$

with corresponding identities also holding for y.[5] But (13.56) and (13.57) are conditions on λ and v, not on $\boldsymbol{A}$. We conclude that the necessary and sufficient condition for matrices to admit non-real right eigenvalues real is that $\boldsymbol{A} \in A_2$:

Lemma 13.3. *The set of* 2×2 *Hermitian matrices* $\boldsymbol{A}$ *for which non-real right eigenvalues is* A_2 *as defined in* (13.38).

Thus, all 2×2 Hermitian matrices which admit non-real right eigenvalues also admit non-real left eigenvalues, and vice versa!

Corollary 13.1. *A* 2×2 *octonionic Hermitian matrix admits right eigenvalues which are not real if and only if it admits left eigenvalues which are not real.*

Turning to the eigenvectors, inserting $p = m$ into (13.52) leads to

$$\overline{x}(ay) = \overline{y}(\overline{a}x) \tag{13.58}$$

and inserting the conditions on a, x, and y now leads to

$$x \cdot y = 0 \tag{13.59}$$

---

[5]This also implies that $a\lambda \cdot x = 0$, that is, x (and y) must be orthogonal to the quaternionic subalgebra generated by λ and a.

just as for left eigenvectors. All right eigenvectors with non-real eigenvalues are hence in V_2, although the converse is false (since right eigenvectors have no real part). Furthermore, not all of the remaining elements of V_2 will be eigenvectors for any given matrix $\boldsymbol{A}$ (since right eigenvectors have no "quaternionic" part).

Putting all of this together, typical solutions of the (right) eigenvalue problem for $\boldsymbol{A}$ as in (13.39) can thus be written as

$$v = n \begin{pmatrix} j \\ k\overline{s} \end{pmatrix}, \qquad \lambda_v = p + q\overline{s}, \tag{13.60}$$

$$w = n \begin{pmatrix} k\overline{s} \\ j \end{pmatrix}, \qquad \lambda_w = p - q\overline{s}, \tag{13.61}$$

where $p, q, n \in \mathbb{R}$ and where

$$s = \cos\theta + k\ell \sin\theta. \tag{13.62}$$

The example given in (13.51) is a special case of the first of (13.61) with $p = q = n = 1$ and $\theta = \pi/2$.

13.3.3   *Further Properties*

We list some further properties of the 2×2 non-real eigenvalue problem without proof. For further details, see [22].

- The eigenvalues of the right eigenvalue problem satisfy the characteristic equation

$$\lambda^2 - \lambda(\operatorname{tr}\boldsymbol{A}) + (\det\boldsymbol{A}) = [\overline{a}, x, y]\frac{(\lambda - p)}{|y|^2} = [a, y, x]\frac{(\lambda - m)}{|x|^2}. \tag{13.63}$$

If the associator $[a, x, y]$ vanishes, then λ satisfies the ordinary characteristic equation, and hence is real (since $\boldsymbol{A}$ is complex Hermitian). Otherwise, comparing real and imaginary parts of the last two terms in (13.63) provides an alternate derivation of $|y| = |x|$, and we recover $p = m$ as expected. Furthermore, since the left-hand side of (13.63) lies in a complex subalgebra of $\mathbb{O}$, so does the right-hand side, and it is then straightforward to solve for λ by considering its real and imaginary parts. The generalized characteristic equation (13.63) then yields the following equation for λ

$$(\operatorname{Re}(\lambda))^2 - \operatorname{Re}(\lambda)(\operatorname{tr}\boldsymbol{A}) + (\det\boldsymbol{A}) = (\operatorname{Im}(\lambda))^2 < 0 \tag{13.64}$$

together with the requirement that

$$\frac{[\overline{a}, x, y]}{|x||y|} = 2\operatorname{Im}(\lambda). \tag{13.65}$$

The explicit form of the eigenvalues given in (13.61) and (13.62) verifies that there are no further restrictions on λ other than (13.56) and (13.64). Furthermore, having shown in the previous subsection that a and x (and therefore also y) are pure imaginary, (13.65) yields an alternate derivation that λ is orthogonal to a, which is (13.56) as well as to x and y, which is (13.57).

- If $Av = \lambda_v v$ with $v^\dagger v = 1$ and $[a, x, y] = 0$, then

$$A = \lambda_v v v^\dagger + \lambda_w w w^\dagger \tag{13.66}$$

  where

$$w = \begin{pmatrix} 0 & 1 \\ 1 & 0 \end{pmatrix} v = \begin{pmatrix} y \\ x \end{pmatrix} \in V_2 \tag{13.67}$$

  and where $\lambda_v = \lambda$ is obtained by solving (13.40) or (13.41), and λ_w is obtained from λ_v by interchanging x and y.

We illustrate this result by returning to the example (13.30), for which we obtain the decomposition

$$\begin{pmatrix} 1 & -i \\ i & 1 \end{pmatrix} = \frac{(1+j)}{2} \begin{pmatrix} 1 \\ k \end{pmatrix} \begin{pmatrix} 1 \\ k \end{pmatrix}^\dagger + \frac{(1-j)}{2} \begin{pmatrix} k \\ 1 \end{pmatrix} \begin{pmatrix} k \\ 1 \end{pmatrix}^\dagger \tag{13.68}$$

where the factor of two is due to the normalization of the eigenvectors. The above construction fails if $[i, x, y] \neq 0$.

- Any $A \in A_2$ with (normalized) $v \in V_2$ such that $Av = v \lambda_v$ can be expanded as

$$A = \lambda_v \left(v v^\dagger \right) + \lambda_w \left(w w^\dagger \right) \tag{13.69}$$

  where w is defined by (13.67) and satisfies $Aw = w \lambda_w$.

As before, we can assume without loss of generality that A is given by (13.39) and that v and w are given by (13.61). Returning to our example (13.51) yields the explicit decomposition

$$\begin{pmatrix} 1 & -i \\ i & 1 \end{pmatrix} = \frac{(1+k\ell)}{2} \left(\begin{pmatrix} j \\ \ell \end{pmatrix} \begin{pmatrix} j \\ \ell \end{pmatrix}^\dagger \right) + \frac{(1-k\ell)}{2} \left(\begin{pmatrix} \ell \\ j \end{pmatrix} \begin{pmatrix} \ell \\ j \end{pmatrix}^\dagger \right). \tag{13.70}$$

  While it is true that

$$\left(v v^\dagger \right) w = v \left(v^\dagger w \right) \tag{13.71}$$

for any v, w related by (13.67) (but not necessarily in V_2), the decomposition (13.69) is surprising because the eigenvalues λ_v, λ_w do *not* commute or associate with the remaining terms. Specifically, although (13.71) is zero here, we have

$$\left(\lambda_v(vv^\dagger)\right) w \neq 0. \tag{13.72}$$

- Any $\boldsymbol{A} \in A_2$ with (normalized) $v \in V_2$ such that $\boldsymbol{A}v = v\lambda_v$ can be expanded as

$$\boldsymbol{A} = (v\lambda_v)\, v^\dagger + (w\lambda_w)\, w^\dagger \tag{13.73}$$

where w is defined by (13.67) and satisfies $\boldsymbol{A}w = w\lambda_w$.

The decomposition (13.73) is less surprising than (13.69) when one realizes that orthogonality in the form

$$\left((v\lambda)\, v^\dagger\right) w = (v\lambda)\, (v^\dagger w) = 0 \tag{13.74}$$

holds for *any* $\lambda \in \mathbb{O}$ and $v, w \in V_2$ satisfying (13.67).

13.4 The 3 × 3 Real Eigenvalue Problem[6]

We now turn to the 3×3 case. It is not immediately obvious that 3×3 octonionic Hermitian matrices have a well-defined characteristic equation. We therefore first review some of the properties of these matrices before turning to the eigenvalue problem. As in the 2×2 case, over the octonions there will be solutions of the eigenvalue problem with eigenvalues which are not real; we consider here only the real eigenvalue problem.

Recall from Section 11.2 that the 3×3 octonionic Hermitian matrices, which we call *Jordan matrices*, form the exceptional Jordan algebra, also known as the Albert algebra, under the Jordan product.

Remarkably, Jordan matrices satisfy the usual characteristic equation [13]

$$\mathcal{A}^3 - (\operatorname{tr}\mathcal{A})\, \mathcal{A}^2 + \sigma(\mathcal{A})\, \mathcal{A} - (\det \mathcal{A})\, \mathcal{I} = 0 \tag{13.75}$$

where $\sigma(\mathcal{A})$ is defined by

$$\sigma(\mathcal{A}) = \frac{1}{2}\left((\operatorname{tr}\mathcal{A})^2 - \operatorname{tr}(\mathcal{A}^2)\right) = \operatorname{tr}(\mathcal{A} * \mathcal{A}) \tag{13.76}$$

and where the determinant of $\mathcal{A}$ was defined abstractly in terms of the Jordan and Freudenthal products in Section 11.2. Concretely, if

$$\mathcal{A} = \begin{pmatrix} p & a & \bar{b} \\ \bar{a} & m & c \\ b & \bar{c} & n \end{pmatrix} \tag{13.77}$$

---

[6]The material in this section is adapted from [21].

with $p, m, n \in \mathbb{R}$ and $a, b, c \in \mathbb{O}$ then

$$\operatorname{tr} \mathcal{A} = p + m + n, \tag{13.78}$$

$$\sigma(\mathcal{A}) = pm + pn + mn - |a|^2 - |b|^2 - |c|^2, \tag{13.79}$$

$$\det \mathcal{A} = pmn + b(ac) + \overline{b(ac)} - n|a|^2 - m|b|^2 - p|c|^2. \tag{13.80}$$

An $n \times n$ Hermitian matrix over any of the normed division algebras can be rewritten as a symmetric $kn \times kn$ real matrix, where k denotes the dimension of the underlying division algebra. It is therefore clear that a 3×3 octonionic Hermitian matrix must have $8 \times 3 = 24$ real eigenvalues [23]. However, as we now show, instead of having (a maximum of) three distinct real eigenvalues, each with multiplicity eight, we show that there are (a maximum of) six distinct real eigenvalues, each with multiplicity four.

The reason for this peculiar property is that, somewhat surprisingly, a (real) eigenvalue λ of a Jordan matrix A does *not* in general satisfy the characteristic equation (13.75).[7] To see this, consider the ordinary eigenvalue equation

$$\mathcal{A}v = \lambda v \tag{13.81}$$

with $\mathcal{A}$ as in (13.77), $\lambda \in \mathbb{R}$, and where

$$v = \begin{pmatrix} x \\ y \\ z \end{pmatrix}. \tag{13.82}$$

Explicit computation yields

$$(\lambda - p)x = ay + \overline{b}z, \tag{13.83}$$

$$(\lambda - m)y = cz + \overline{a}x, \tag{13.84}$$

$$(\lambda - n)z = bx + \overline{c}y, \tag{13.85}$$

so that

$$(\lambda - p)(\lambda - m)y = (\lambda - p)(cz + \overline{a}x) = (\lambda - p)cz + \overline{a}(ay + \overline{b}z) \tag{13.86}$$

which implies

$$\left[(\lambda - p)(\lambda - m) - |a|^2\right] y = \overline{a}(\overline{b}z) + (\lambda - p)cz. \tag{13.87}$$

---

[7]Ogievetskiĭ [24] constructed a 6th order polynomial satisfied by the real eigenvalues, which he called the characteristic equation. This polynomial is presumably equivalent to the modified characteristic equations (for both values of r) given below.

Assume first that $\lambda \neq p$. Using (13.83) and (13.87) in (13.85) leads to

$$
\begin{aligned}
&\left[(\lambda - p)(\lambda - m) - |a|^2\right] (\lambda - p)(\lambda - n)z \\
&= \left[(\lambda - p)(\lambda - m) - |a|^2\right] (\lambda - p)(bx + \bar{c}y) \\
&= \left[(\lambda - p)(\lambda - m) - |a|^2\right] b(ay + \bar{b}z) + (\lambda - p)\bar{c}\left[\bar{a}(\bar{b}z) + (\lambda - p)cz\right] \\
&= b\left[a\left(\bar{a}(\bar{b}z) + (\lambda - p)cz\right)\right] + \left[(\lambda - p)(\lambda - m) - |a|^2\right] b(\bar{b}z) \\
&\quad + (\lambda - p)\bar{c}\left[\bar{a}(\bar{b}z) + (\lambda - p)cz\right] \\
&= (\lambda - p)\left[(\lambda - m)|b|^2 z + (\lambda - p)|c|^2 z + b(a(cz)) + \bar{c}\left(\bar{a}(\bar{b}z)\right)\right]. \quad (13.88)
\end{aligned}
$$

Expanding this out and comparing with (13.78)–(13.80) results finally in

$$
\begin{aligned}
\left[\det(\lambda I - A)\right]z &= \left[\lambda^3 - (\operatorname{tr} A)\lambda^2 + \sigma(A)\lambda - \det A\right]z \\
&= b(a(cz)) + \bar{c}\left(\bar{a}(\bar{b}z)\right) - \left[b(ac) + (\bar{c}\,\bar{a})\bar{b}\right]z. \quad (13.89)
\end{aligned}
$$

Now consider the case $\lambda = p$. We still have (13.87), which here takes the form

$$
-|a|^2 y = \bar{a}(\bar{b}z). \quad (13.90)
$$

Inserting (13.90) into (13.84), we can solve for x, obtaining

$$
-|a|^2 x = a(cz) + (p - m)\bar{b}z. \quad (13.91)
$$

Finally, inserting (13.90) and (13.91) in (13.85) yields

$$
-\left(|a|^2(p - n) + |b|^2(p - m)\right)z = b(a(cz)) + \bar{c}\left(\bar{a}(\bar{b}z)\right). \quad (13.92)
$$

Comparing with (13.78)–(13.80) and using $\lambda = p$, we see that (13.89) still holds, and thus holds in general.

If a, b, c, and z associate, the right-hand side of (13.89) vanishes, and λ does indeed satisfy the characteristic equation (13.75); this will not happen in general. However, since the left-hand side of (13.89) is a real multiple of z, this must also be true of the right-hand side, so that

$$
b(a(cz)) + \bar{c}\left(\bar{a}(\bar{b}z)\right) - \left[b(ac) + (\bar{c}\,\bar{a})\bar{b}\right]z = rz \quad (13.93)
$$

for some $r \in \mathbb{R}$, which can be solved to yield a quadratic equation for r as well as constraints on z.

Lemma 13.4. *The real eigenvalues of the* 3×3 *octonionic Hermitian matrix* $\mathcal{A}$ *satisfy the modified characteristic equation*

$$
\det(\lambda I - \mathcal{A}) = \lambda^3 - (\operatorname{tr} \mathcal{A})\lambda^2 + \sigma(\mathcal{A})\lambda - \det \mathcal{A} = r \quad (13.94)
$$

where r is either of the two roots of

$$r^2 + 4\Phi(a, b, c)\, r - |[a, b, c]|^2 = 0 \qquad (13.95)$$

with a, b, c as defined by (13.77) and where

$$\Phi(a, b, c) = \frac{1}{2}\,\mathrm{Re}\,([a, \bar{b}]c). \qquad (13.96)$$

Proof. These results were obtained using *Mathematica* to solve (13.93) by brute force for real r and octonionic z given generic octonions a, b, c [25]. □

Furthermore, provided that $[a, b, c] \neq 0$, each of x, y, and z can be shown to admit an expansion in terms of four real parameters.

Corollary 13.2. *With $\mathcal{A}$ and r as above, and assuming $[a, b, c] \neq 0$,*

$$z = (\alpha a + \beta b + \gamma c + \delta)\left(1 + \frac{[a, b, c]\, r}{|[a, b, c]|^2}\right) \qquad (13.97)$$

with $\alpha, \beta, \gamma, \delta \in \mathbb{R}$. Similar expansions hold for x and y.

The real parameters $\alpha, \beta, \gamma, \delta$ may be freely specified for one (nonzero) component, say z; the remaining components x, y have a similar form which is then fully determined by (13.83)–(13.85).

Corollary 13.3. *The real eigenvalues of $\overline{\mathcal{A}}$ are the same as those of $\mathcal{A}$.*

Proof. Direct computation (or (13.78)–(13.80)) shows that

$$\det \overline{\mathcal{A}} = \det \mathcal{A} - 4\Phi(a, b, c). \qquad (13.98)$$

But $-4\Phi(a, b, c)$ is precisely the sum of the roots of (13.95), and replacing $\mathcal{A}$ by $\overline{\mathcal{A}}$ merely flips the sign of r, that is $r[\overline{\mathcal{A}}] = -r[\mathcal{A}]$. Thus, the two possible values of $\det \mathcal{A} + r[\mathcal{A}]$ are precisely the same as those for $\det \overline{\mathcal{A}} + r[\overline{\mathcal{A}}]$. Since $\mathrm{tr}\,\overline{\mathcal{A}} = \mathrm{tr}\,\mathcal{A}$ and $\sigma(\overline{\mathcal{A}}) = \sigma(\mathcal{A})$, (13.94) is unchanged. □

The solutions of (13.94) are real, since the corresponding 24×24 real symmetric matrix has 24 real eigenvalues. We will refer to the three real solutions of (13.94) corresponding to a single value of r as a *family* of eigenvalues of $\mathcal{A}$. There are thus two families of real eigenvalues, each corresponding to four independent (over $\mathbb{R}$) eigenvectors.

We note several intriguing properties of these results. If $\mathcal{A}$ is in fact complex, then the only solution of (13.95) is $r = 0$, and we recover the usual characteristic equation with a unique set of three (real) eigenvalues.

If $\mathcal{A}$ is quaternionic, then one solution of (13.95) is $r = 0$, leading to the standard set of three real eigenvalues and their corresponding quaternionic eigenvectors. However, unless a, b, c involve only two independent imaginary quaternionic directions (in which case $\Phi(a, b, c) = 0 = [a, b, c]$), there will also be a nonzero solution for r, leading to a second set of three real eigenvalues. From the preceding corollary, we see that this second set of eigenvalues consists precisely of the usual ($r = 0$) eigenvalues of $\overline{\mathcal{A}}$! Furthermore, since

$$\mathcal{A}(\ell v) = \ell(\overline{\mathcal{A}} v), \qquad (13.99)$$

if the components of $\mathcal{A}$ and v are in $\mathbb{H}$ and for imaginary $\ell \in \mathbb{O}$ orthogonal to $\mathbb{H}$, the eigenvectors of $\mathcal{A}$ corresponding to $r \neq 0$ are precisely ℓ times the quaternionic ($r = 0$) eigenvectors of $\overline{\mathcal{A}}$. In this sense, the octonionic eigenvalue problem for quaternionic $\mathcal{A}$ is equivalent to the quaternionic eigenvalue problem for both $\mathcal{A}$ and $\overline{\mathcal{A}}$ together. Finally, if $\mathcal{A}$ is octonionic (so that in particular $[a, b, c] \neq 0$), then there are two distinct solutions for r, and hence two different sets of real eigenvalues, with corresponding eigenvectors. Note that if $\det \mathcal{A} = 0 \neq [a, b, c]$ then all of the eigenvalues of $\mathcal{A}$ will be nonzero!

The final surprise lies with the orthogonality condition for eigenvectors v, w corresponding to different eigenvalues. As in the 2×2 case considered in Section 13.2, it is *not* true that $v^\dagger w = 0$, although the real part of this expression does vanish, that is,

$$v^\dagger w + w^\dagger v = 0. \qquad (13.100)$$

However, just as in the 2×2 case, what is needed to ensure a decomposition of the form

$$\mathcal{A} = \sum_{m=1}^{3} \lambda_m v_m v_m^\dagger \qquad (13.101)$$

is (13.27), and a lengthy, direct computation verifies that (13.27) holds *provided that* both eigenvectors correspond to the same value of r.

Lemma 13.5. *If v and w are eigenvectors of the 3×3 octonionic Hermitian matrix $\mathcal{A}$ corresponding to different real eigenvalues in the same family (same r value), then v and w are mutually orthogonal in the sense of (13.27), that is, $(vv^\dagger)w = 0$.*

Proof. The modified characteristic equation (13.94) can be used to eliminate cubic and higher powers of λ from any expression. Furthermore, given two distinct eigenvalues $\lambda_1 \neq \lambda_2$, subtracting the two versions of (13.94) and factoring the result leads to the equation

$$(\lambda_1^2 + \lambda_1\lambda_2 + \lambda_2^2) - \operatorname{tr}\mathcal{A}(\lambda_1 + \lambda_2) + \sigma(\mathcal{A}) = 0 \qquad (13.102)$$

thus eliminating quadratic terms in one of the eigenvalues. The result now follows by (lengthy!) computation.[8] □

For Jordan matrices, we thus obtain *two* decompositions of the form (13.101), corresponding to the two sets of real eigenvalues. For each, the eigenvectors are fixed up to orthogonal transformations which preserve the form (13.97) of z.

Theorem 13.1. *Let $\mathcal{A}$ be a 3×3 octonionic Hermitian matrix. Then $\mathcal{A}$ can be expanded as in (13.101), where $\{v_1, v_2, v_3\}$ are orthonormal (as per (13.27)) eigenvectors of $\mathcal{A}$ corresponding to the real eigenvalues λ_m, which belong to the same family (same r value).*

Proof. Fix a family of real eigenvalues of $\mathcal{A}$ by fixing r. If the eigenvalues are distinct, then the previous lemma guarantees the existence of orthonormal eigenvectors, which are also eigenvectors of the decomposition (13.101) with the same eigenvalues, and the result follows.

If the eigenvalues are the same, the family consists of a single real eigenvalue λ with multiplicity three. Then $\operatorname{tr}(\mathcal{A}) = 3\lambda$ and $\sigma(\mathcal{A}) = 3\lambda^2$. Writing out these two equations in terms of the components (13.77) of $\mathcal{A}$, and inserting the first into the second, results in a quadratic equation for λ; the discriminant D of this equation satisfies $D \leq 0$. But λ is assumed to be real, which forces $D = 0$, which in turn forces $\mathcal{A}$ to be a multiple of the identity matrix, for which the result holds.

The remaining case is when one eigenvalue, say μ, has multiplicity two and one has multiplicity one. Letting v be a (normalized) eigenvector with eigenvalue μ, consider the matrix

$$X = \mathcal{A} - \alpha\,vv^\dagger \qquad (13.103)$$

with $\alpha \in \mathbb{R}$. For most values of α, X will have three distinct real eigenvalues, whose eigenvectors will be orthogonal by the previous theorem. But this means that eigenvectors of X are also eigenvectors of $\mathcal{A}$; the required

---

[8]We used *Mathematica* to implement these simplifications in a brute force verification of (13.27) in this context, which ran for six hours on a SUN Sparc20 with 224 Mb of RAM [25]. An amusing footnote to this story is that the subsequent version of *Mathematica* was unable to finish this computation on the same hardware, although later versions (on better hardware) can reproduce the computation much more quickly.

decomposition of $\mathcal{A}$ is obtained from that of X simply by solving (13.103) for $\mathcal{A}$.   □

Note in particular that for some quaternionic matrices with determinant equal to zero, one and only one of these two decompositions will contain the eigenvalue zero.

In the 2×2 case,

$$(vv^\dagger)(vv^\dagger) = (v^\dagger v)(vv^\dagger) \tag{13.104}$$

which tells us that, for normalized v, $vv^\dagger$ squares to itself, and hence is idempotent. The decomposition analogous to (13.101) in the 2×2 case is thus an idempotent decomposition. But (13.104) fails in the 3×3 case, so that the decomposition in Theorem 13.1 is therefore *not* an idempotent decomposition.

It is nevertheless straightforward to show that if u, v, and w are orthonormal in the sense of (13.27) then

$$uu^\dagger + vv^\dagger + ww^\dagger = \mathcal{I} \tag{13.105}$$

since the left-hand side has eigenvalue 1 with multiplicity three. This permits us to view $\{u, v, w\}$ as a basis of $\mathbb{O}^3$ in the following sense

Lemma 13.6. *Let $u, v, w \in \mathbb{O}^3$ be orthonormal in the sense of (13.27) and let g be any vector in $\mathbb{O}^3$. Then*

$$g = (uu^\dagger)\, g + (vv^\dagger)\, g + (ww^\dagger)\, g. \tag{13.106}$$

Proof. This result follows immediately from (13.105).   □

However, another consequence of the failure of (13.104) in the 3×3 case is that the Gram–Schmidt orthogonalization procedure no longer works. It appears to be fortuitous that we are nevertheless able to find orthonormal eigenvectors in the 3×3 case with repeated eigenvalues; we suspect that this might fail in general, perhaps already in the 4×4 case with an eigenvalue of multiplicity three.

We can relate our notion of orthonormality to the usual one by noting that n vectors in $\mathbb{O}^n$ which are orthonormal in the sense (13.27) satisfy

$$vv^\dagger + \cdots + ww^\dagger = \mathcal{I}. \tag{13.107}$$

If we define a matrix $\mathcal{U}$ whose columns are just $v, \ldots, w$, then this statement is equivalent to

$$\mathcal{U}\mathcal{U}^\dagger = \mathcal{I}. \tag{13.108}$$

Over the quaternions, left matrix inverses are the same as right matrix inverses, and we would also have

$$\mathcal{U}^\dagger \mathcal{U} = \mathcal{I} \tag{13.109}$$

or equivalently

$$v^\dagger v = 1 = \cdots = w^\dagger w, \qquad v^\dagger w = 0 = \cdots \tag{13.110}$$

which is just the standard notion of orthogonality. These two notions of orthogonality fail to be equivalent over the octonions; we have been led to view the former as more fundamental.

We can now rewrite the eigenvalue equation in the form

$$\mathcal{A}\mathcal{U} = \mathcal{U}\mathcal{D} \tag{13.111}$$

where $\mathcal{D}$ is a diagonal matrix whose entries are the real eigenvalues.[9] Multiplying (13.111) on the left by $\mathcal{U}^\dagger$ yields

$$\mathcal{U}^\dagger(\mathcal{A}\mathcal{U}) = \mathcal{U}^\dagger(\mathcal{U}\mathcal{D}) = (\mathcal{U}^\dagger\mathcal{U})\mathcal{D} \tag{13.112}$$

(since $\mathcal{D}$ is real), but this does not lead to a diagonalization of $\mathcal{A}$ since, as noted above, $\mathcal{U}^\dagger\mathcal{U}$ is not in general equal to the identity matrix. However, Theorem 13.1 can be rewritten as

$$\mathcal{A} = \mathcal{U}\mathcal{D}\mathcal{U}^\dagger \tag{13.113}$$

so that in this sense $\mathcal{A}$ is diagonalizable. Furthermore, multiplication of (13.111) on the right by $\mathcal{U}^\dagger$ shows that

$$(\mathcal{A}\mathcal{U})\mathcal{U}^\dagger = (\mathcal{U}\mathcal{D})\mathcal{U}^\dagger = \mathcal{A} = \mathcal{A}(\mathcal{U}\mathcal{U}^\dagger) \tag{13.114}$$

and this assertion of associativity can be taken as a restatement of Theorem 13.1. In the 2×2 case, this associativity holds for a single eigenvector v, that is,

$$(\boldsymbol{A}v)v^\dagger = \boldsymbol{A}(vv^\dagger) \tag{13.115}$$

which leads to the elegant one-line derivation

$$\boldsymbol{A} = \boldsymbol{A}\left(\sum_{m=1}^{2} v_m v_m^\dagger\right) = \sum_{m=1}^{2} \boldsymbol{A}(v_m v_m^\dagger) = \sum_{m=1}^{2} (\boldsymbol{A}v_m)v_m^\dagger = \sum_{m=1}^{2} \lambda_m v_m v_m^\dagger. \tag{13.116}$$

However, (13.115) fails in the 3×3 case, and we are unaware of a correspondingly elegant proof of Theorem 13.1.

Our original proof of the 3×3 orthogonality result (13.5) used *Mathematica* to explicitly perform a horrendous, but exact, algebraic computation. Although Okubo [26] did later give an analytic proof of this result, the *Mathematica* computation nevertheless establishes a result which would otherwise have remained merely a conjecture. This is a good example of being able to use the computer to verify one's intuition when it may not be possible to do so otherwise. This issue is further discussed in [25].

---

[9] Since the eigenvalues are real, it doesn't matter if we put them on the right.

13.5   The 3 × 3 Non-real Eigenvalue Problem

Without being able to solve (some version of) the characteristic equation
in the 3×3 case, it is not possible in general to determine all the (non-
real) eigenvalues of a given Hermitian octonionic matrix. It is therefore
instructive to consider several explicit examples.

13.5.1   *Example 1*

Consider the matrix

$$\mathcal{B} = \begin{pmatrix} p & iq & kqs \\ -iq & p & jq \\ -kqs & -jq & p \end{pmatrix} \tag{13.117}$$

where

$$s = \cos\theta + k\ell \sin\theta. \tag{13.118}$$

Note that $\mathcal{B}$ is quaternionic if $\theta = 0$.

The real eigenvalues of $\mathcal{B}$, and corresponding orthonormal bases of eigenvectors, were given in [25]. But $\mathcal{B}$ also admits eigenvectors with eigenvalues
which are not real. For instance:

$$\lambda_{\widehat{u}} = p \pm q\bar{s}, \quad \widehat{u}_\pm = \begin{pmatrix} i \\ 0 \\ j \end{pmatrix} S_\pm, \tag{13.119}$$

$$\lambda_{\widehat{v}} = p \pm q\bar{s}, \quad \widehat{v}_\pm = \begin{pmatrix} j \\ 2ks \\ i \end{pmatrix} S_\pm, \tag{13.120}$$

$$\lambda_{\widehat{w}} = p \mp 2q\bar{s}, \quad \widehat{w}_\pm = \begin{pmatrix} j \\ -ks \\ i \end{pmatrix} S_\pm, \tag{13.121}$$

where

$$S_\pm = \begin{cases} -k\ell, \\ 1. \end{cases} \tag{13.122}$$

These eigenvectors and eigenvalues reduce to the ones given in [25] when
$\theta \to 0$. Somewhat surprisingly, these eigenvectors (when normalized) yield
a decomposition of the form (13.101). Remarkably, they also yield a decomposition of the form

$$\mathcal{B} = \sum_{\alpha=1}^{3} (v_\alpha \lambda_\alpha) \, v_\alpha^\dagger. \tag{13.123}$$

13.5.2 Example 2

A related example is given by the matrix

$$\widehat{\mathcal{B}} = \begin{pmatrix} p & qi & \frac{q}{2}ks \\ -qi & p & \frac{q}{2}j \\ -\frac{q}{2}ks & -\frac{q}{2}jp \end{pmatrix} \tag{13.124}$$

with s again given by (13.118). We choose θ such that

$$s = \frac{\sqrt{5}}{3} - \frac{2}{3}k\ell \tag{13.125}$$

resulting in

$$\widehat{\mathcal{B}} = \begin{pmatrix} p & qi & \frac{q}{6}(\sqrt{5}k+2l) \\ -qi & p & \frac{q}{2}j \\ -\frac{q}{6}(\sqrt{5}k+2l) & -\frac{q}{2}j & p \end{pmatrix}. \tag{13.126}$$

The two families of real eigenvalues of $\widehat{\mathcal{B}}$ turn out to be $\{p\pm q, p\mp\frac{q}{2}(1+\sqrt{3}),$ $p\mp\frac{q}{2}(1-\frac{\sqrt{3}}{2})\}$. Some eigenvectors for $\widehat{\mathcal{B}}$ corresponding to eigenvalues which are not real are

$$\lambda_{u_1} = \left(p+\frac{\sqrt{5}}{2}q\right) - \frac{q}{2}k\ell, \quad u_1 = \begin{pmatrix} 3k \\ \sqrt{5}j - 2i\ell \\ 1 + \sqrt{5}k\ell \end{pmatrix}, \tag{13.127}$$

$$\lambda_{u_2} = \left(p+\frac{\sqrt{5}}{2}q\right) + \frac{q}{2}k\ell, \quad u_2 = \begin{pmatrix} \sqrt{5}k + 2\ell \\ 3j \\ \sqrt{5} - k\ell \end{pmatrix}, \tag{13.128}$$

$$\lambda_{v_1} = \left(p-\frac{\sqrt{5}}{3}q\right) + \frac{2q}{3}k\ell, \quad v_1 = \begin{pmatrix} \sqrt{5}j - 2i\ell \\ 3k \\ 0 \end{pmatrix}, \tag{13.129}$$

$$\lambda_{v_2} = \left(p-\frac{\sqrt{5}}{3}q\right) - \frac{2q}{3}k\ell, \quad v_2 = \begin{pmatrix} 3j \\ \sqrt{5}k + 2\ell \\ 0 \end{pmatrix}, \tag{13.130}$$

$$\lambda_{w_1} = \left(p-\frac{\sqrt{5}}{6}q\right) - \frac{q}{6}k\ell, \quad w_1 = \begin{pmatrix} 3k \\ \sqrt{5}j - 2i\ell \\ -7 - \sqrt{5}k\ell \end{pmatrix}, \tag{13.131}$$

$$\lambda_{w_2} = \left(p-\frac{\sqrt{5}}{6}q\right) + \frac{q}{6}k\ell, \quad w_2 = \begin{pmatrix} \sqrt{5}k + 2l \\ 3j \\ -3\sqrt{5} - 3k\ell \end{pmatrix}. \tag{13.132}$$

However, we have been unable to find any decompositions of $\widehat{\mathcal{B}}$ involving these vectors. It is intriguing that, for instance, v_1 is orthogonal to both u_1 and w_1 (in the sense of (13.27), but that u_1 and w_1 are not orthogonal. In fact, we have shown using *Mathematica* that there is *no* eigenvector triple containing w_1 which is orthogonal in the sense of (13.27). Unless w_1 is special in some as yet to be determined sense, we are forced to conclude that neither (13.27) nor (13.101) are generally true for eigenvectors whose eigenvalues are not real. It is curious, however, that the sum of the squares (outer products) of all *six* of these (normalized) vectors is indeed (twice) the identity!

13.5.3   *Example 3*

In all of the examples considered so far, the eigenvalues have been in the complex subalgebra of $\mathbb{O}$ determined by the associator $[a, b, c]$ (with a, b, c as in (13.77)). We now give an example for which this is not the case.

Consider

$$
\mathcal{C} = \begin{pmatrix} p & iq & -q(j - i\ell - j\ell) \\ -iq & p & q(1 + k + l) \\ q(j - i\ell - j\ell) & -q(1 - k - l)p & \end{pmatrix} \tag{13.133}
$$

which admits an eigenvector

$$
v = \begin{pmatrix} j \\ l \\ 0 \end{pmatrix} \tag{13.134}
$$

with eigenvalue

$$
\lambda_v = p + q\, l k. \tag{13.135}
$$

However, the associator takes the form

$$
\frac{[a, b, c]}{q^3} = [i, (j - i\ell - j\ell), (1 + k + l)] = 2(l - k). \tag{13.136}
$$

Further discussion of these examples appears in [27].

13.6   The Jordan Eigenvalue Problem[10]

In Sections 13.2–13.5, we considered both the *left* and *right* eigenvalue problems for 2×2 and 3×3 octonionic Hermitian matrices. In the 3×3

---

[10]The material in this section is adapted from [28].

octonionic case, even if we assume that the eigenvalues are real, the eigenvalue problem does not quite behave as expected. For this case, there are six, rather than three, real eigenvalues [24], which come in two independent families, each consisting of three real eigenvalues which satisfy a modified characteristic equation rather than the usual one. Furthermore, the corresponding eigenvectors are not orthogonal in the usual sense, but do satisfy a generalized notion of orthogonality (see also [25, 26]). Finally, all such matrices admit a decomposition in terms of (the "squares" of) orthonormal eigenvectors. However, due to associativity problems, these matrices are *not* idempotents (matrices which square to themselves).

We describe here a related eigenvalue problem for 3×3 Hermitian octonionic matrices which does have the standard properties: There are three real eigenvalues, which solve the usual characteristic equation, and which lead to a decomposition in terms of orthogonal "eigenvectors" which are indeed (primitive) idempotents.

We consider the *eigenmatrix* problem

$$\mathcal{A} \circ \mathcal{V} = \lambda \mathcal{V} \tag{13.137}$$

where $\mathcal{V}$ is itself an octonionic Hermitian matrix and $\circ$ denotes the *Jordan product* [29, 30]

$$\mathcal{A} \circ \mathcal{B} = \frac{1}{2} (\mathcal{A}\mathcal{B} + \mathcal{B}\mathcal{A}) \tag{13.138}$$

which is commutative but not associative. We further restrict $\mathcal{V}$ to be a (primitive) idempotent; as discussed below, this ensures that the Jordan eigenvalue problem (13.137) reduces to the traditional eigenvalue problem (13.9) in the non-octonionic cases. Since $\mathcal{A}$ and $\mathcal{V}$ are Jordan matrices, the left-hand side of (13.137) is Hermitian, which forces λ to be real.

Suppose first that $\mathcal{A}$ is diagonal. Then the diagonal elements p, m, n are clearly eigenvalues, with obvious diagonal eigenmatrices. But there are also other eigenvalues! For example,

$$\begin{pmatrix} p & 0 & 0 \\ 0 & m & 0 \\ 0 & 0 & n \end{pmatrix} \circ \begin{pmatrix} 0 & 1 & 0 \\ 1 & 0 & 0 \\ 0 & 0 & 0 \end{pmatrix} = \frac{p+m}{2} \begin{pmatrix} 0 & 1 & 0 \\ 1 & 0 & 0 \\ 0 & 0 & 0 \end{pmatrix} \tag{13.139}$$

so the averages $(p+m)/2$, $(m+n)/2$, $(n+p)/2$ are also eigenvalues. However, the corresponding eigenmatrices—which are related to Peirce decompositions [31, 32]—have only zeros on the diagonal, and cannot therefore square to themselves. To exclude this case, we restrict $\mathcal{V}$ in (13.137) to the Cayley plane, which ensures that the eigenmatrices are primitive idempotents; as we will see, they really do correspond to "eigenvectors" v. Recall that this condition forces the components of $\mathcal{V}$ to lie in a quaternionic subalgebra of $\mathbb{O}$ (which depends on $\mathcal{V}$) even though the components of $\mathcal{A}$ may not.

Next consider the characteristic equation (13.75) in the form

$$-\det(\mathcal{A} - \lambda \mathcal{I}) = \lambda^3 - (\operatorname{tr} \mathcal{A}) \lambda^2 + \sigma(\mathcal{A}) \lambda - (\det \mathcal{A}) \mathcal{I} = 0. \qquad (13.140)$$

It is not at first obvious that all solutions λ of (13.140) are real. To see that this is indeed the case, we note that $\mathcal{A}$ can be rewritten as a 24×24 real symmetric matrix, whose eigenvalues are of course real. However, as discussed in Section 13.1, these latter eigenvalues do *not* satisfy the characteristic equation (13.140)! Rather, they satisfy a modified characteristic equation of the form

$$\det(\mathcal{A} - \lambda \mathcal{I}) + r = 0 \qquad (13.141)$$

where r is either of the roots of a quadratic equation which depends on $\mathcal{A}$, namely (13.95). As shown explicitly using *Mathematica* in Figure 5 of [25], not only are these roots real, but they have opposite signs (or at least one is zero). But, as can be seen immediately using elementary graphing techniques, if the cubic equation (13.141) has three real roots for both a positive and a negative value of r, it also has three real roots for all values of r in between, including $r = 0$. This argument shows that (13.140) does indeed have three real roots.

Alternatively, since F_4 preserves both the determinant and the trace (and therefore also σ) [13,33], it leaves the characteristic equation invariant. Since F_4 can be used to diagonalize $\mathcal{A}$ [13, 28, 33] (see also Section 13.7), and since the resulting diagonal elements clearly satisfy the characteristic equation, we have another, indirect, proof that the characteristic equation has three real roots. Furthermore, this argument shows that these roots correspond precisely to the three real eigenvalues whose eigenmatrices lie in the Cayley plane. We therefore reserve the word "eigenvalue" for the three solutions of the characteristic equation (13.140), explicitly excluding their averages. The above argument shows that these eigenvalues correspond to solutions $\mathcal{V}$ of (13.137) which lie in the Cayley plane; we will verify this explicitly below.

Restricting the eigenvalues in this way corresponds to the traditional eigenvalue problem in the following sense. If the components of $\mathcal{A}$ and $v \neq 0$ lie in a quaternionic subalgebra of the octonions, then the Jordan eigenvalue problem (13.137) together with the restriction to the Cayley plane becomes

$$\mathcal{A}\, vv^{\dagger} + vv^{\dagger} \mathcal{A} = 2\lambda\, vv^{\dagger}. \qquad (13.142)$$

Multiplying (13.142) on the right by v and simplifying the result using the trace of (13.142) leads immediately to $Av = \lambda v$ (with $\lambda \in \mathbb{R}$), that is, the Jordan eigenvalue equation implies the ordinary eigenvalue equation in this context. Since the converse is immediate, the Jordan eigenvalue problem (13.137) (with $\mathcal{V}$ restricted to the Cayley plane but $\mathcal{A}$ octonionic) is seen to be a reasonable generalization of the ordinary eigenvalue problem.

We now show how to construct eigenmatrices $\mathcal{V}$ of (13.137), restricted to lie in the Cayley plane, and with real eigenvalues λ satisfying the characteristic equation (13.140). From the definition of the determinant, we have for real λ that satisfying (13.140)

$$0 = \det(\mathcal{A} - \lambda \mathcal{I}) = (\mathcal{A} - \lambda \mathcal{I}) \circ ((\mathcal{A} - \lambda \mathcal{I}) * (\mathcal{A} - \lambda \mathcal{I})). \qquad (13.143)$$

Thus, setting

$$\mathcal{Q}_\lambda = (\mathcal{A} - \lambda \mathcal{I}) * (\mathcal{A} - \lambda \mathcal{I}) \qquad (13.144)$$

we have

$$(\mathcal{A} - \lambda \mathcal{I}) \circ \mathcal{Q}_\lambda = 0 \qquad (13.145)$$

so that $\mathcal{Q}_\lambda$ is a solution of (13.137).

Due to the identity

$$(\mathcal{X} * \mathcal{X}) * (\mathcal{X} * \mathcal{X}) = (\det \mathcal{X}) \, \mathcal{X} \qquad (13.146)$$

we have

$$\mathcal{Q}_\lambda * \mathcal{Q}_\lambda = 0. \qquad (13.147)$$

If $\mathcal{Q}_\lambda \neq 0$, we can renormalize $\mathcal{Q}_\lambda$ by defining

$$\mathcal{P}_\lambda = \frac{\mathcal{Q}_\lambda}{\operatorname{tr}(\mathcal{Q}_\lambda)}. \qquad (13.148)$$

Each resulting $\mathcal{P}_\lambda$ is in the Cayley plane, and is hence a primitive idempotent. Due to (13.147) and (12.69), we can write

$$\mathcal{P}_\lambda = v_\lambda v_\lambda^\dagger \qquad (13.149)$$

and we call v_λ the (generalized) eigenvector of $\mathcal{A}$ with eigenvalue λ. Note that v_λ does *not* in general satisfy either (13.1) or (13.9). Rather, we have

$$\mathcal{A} \circ v_\lambda v_\lambda^\dagger = \lambda \, v_\lambda v_\lambda^\dagger \qquad (13.150)$$

as well as

$$v_\lambda^\dagger v_\lambda = 1. \qquad (13.151)$$

Writing out all the terms and using the identities

$$\widetilde{\mathcal{X}} = \mathcal{X} - \text{tr}\,(\mathcal{X})\,\mathcal{I} = -2\mathcal{I} * \mathcal{X}, \tag{13.152}$$

$$(\widetilde{\mathcal{X}} \circ \mathcal{X}) \circ (\mathcal{X} * \mathcal{X}) = (\det \mathcal{X})\,\widetilde{\mathcal{X}}, \tag{13.153}$$

one computes directly that

$$\mathcal{Q}_\lambda \circ (\mathcal{A} \circ \mathcal{Q}_\mu) = (\mathcal{Q}_\lambda \circ \mathcal{A}) \circ \mathcal{Q}_\mu. \tag{13.154}$$

If λ, μ are solutions of the characteristic equation (13.140), then using (13.145) leads to

$$\mu\,(\mathcal{Q}_\lambda \circ \mathcal{Q}\mu) = \lambda\,(\mathcal{Q}_\lambda \circ \mathcal{Q}\mu). \tag{13.155}$$

If we now assume $\lambda \neq \mu$ and $\mathcal{Q}_\lambda \neq 0 \neq \mathcal{Q}_\mu$, this computation shows that eigenmatrices corresponding to different eigenvalues are orthogonal in the sense

$$\mathcal{P}_\lambda \circ \mathcal{P}_\mu = 0 \tag{13.156}$$

where we have normalized the eigenmatrices.

We now turn to the case $\mathcal{Q}_\lambda = 0$. We have first that

$$\text{tr}\,(\mathcal{Q}_\lambda) = \text{tr}\,((\mathcal{A} - \lambda\mathcal{I}) * (\mathcal{A} - \lambda\mathcal{I})) = \sigma(\mathcal{A} - \lambda\mathcal{I}). \tag{13.157}$$

Denoting the three real solutions of the characteristic equation (13.140) by λ, μ, ν, so that

$$\text{tr}\,\mathcal{A} \;=\; \lambda + \mu + \nu, \tag{13.158}$$

$$\sigma(\mathcal{A}) \;=\; \lambda(\mu + \nu) + \mu\nu, \tag{13.159}$$

we then have

$$\sigma(\mathcal{A} - \lambda\mathcal{I}) = \sigma(\mathcal{A}) - 2\lambda\,\text{tr}\,\mathcal{A} + 3\lambda^2 = (\lambda - \mu)(\lambda - \nu). \tag{13.160}$$

But by (13.147) and (12.78), $\mathcal{Q}_\lambda = 0$ if and only if $\text{tr}\,(\mathcal{Q}_\lambda) = 0$. Using (13.157) and (13.160), we therefore see that $\mathcal{Q}_\lambda = 0$ if and only if λ is a solution of (13.140) of multiplicity greater than one. We will return to this case below.

Putting this all together, if there are no repeated solutions of the characteristic equation (13.140), then the eigenmatrix problem leads to the decomposition

$$\mathcal{A} = \sum_{i=1}^{3} \lambda_i \mathcal{P}_{\lambda_i} \tag{13.161}$$

in terms of orthogonal primitive idempotents, which expresses each Jordan matrix $\mathcal{A}$ as a sum of squares of *quaternionic* columns.[11] We emphasize that the components of the eigenmatrices $\mathcal{P}_{\lambda_i}$ need not lie in the same quaternionic subalgebra, and that $\mathcal{A}$ is octonionic. Nonetheless, it is remarkable that $\mathcal{A}$ admits a decomposition in terms of matrices which are, individually, quaternionic.

We now return to the case $\mathcal{Q}_\lambda = 0$, corresponding to repeated eigenvalues. If λ is a solution of the characteristic equation (13.140) of multiplicity three, then $\operatorname{tr}\mathcal{A} = 3\lambda$ and $\sigma(\mathcal{A}) = 3\lambda^2$. As shown in [21] in a different context, or using an argument along the lines of footnote 11, this forces $\mathcal{A} = \lambda\mathcal{I}$, which has a trivial decomposition into orthonormal primitive idempotents. We are left with the case of multiplicity two, corresponding to $\mathcal{A} \neq \lambda\mathcal{I}$ and $\mathcal{Q}_\lambda = 0$.

Since $\mathcal{Q}_\lambda = 0$, $\mathcal{A} - \lambda\mathcal{I}$ is (up to normalization) in the Cayley plane, and we have

$$\mathcal{A} - \lambda\mathcal{I} = \pm ww^\dagger \tag{13.162}$$

with the components of w in some quaternionic subalgebra of $\mathbb{O}$. While $ww^\dagger$ is indeed an eigenmatrix of $\mathcal{A}$, it has eigenvalue $\mu = \operatorname{tr}(\mathcal{A}) - 2\lambda \neq \lambda$. However, it is straightforward to construct a vector v orthogonal to w in a suitable sense. For instance, if

$$w = \begin{pmatrix} x \\ y \\ r \end{pmatrix} \tag{13.163}$$

with $r \in \mathbb{R}$, then choosing

$$v = \begin{pmatrix} |y|^2 \\ -y\overline{x} \\ 0 \end{pmatrix} \tag{13.164}$$

leads to

$$vv^\dagger \circ ww^\dagger = 0 \tag{13.165}$$

and only minor modifications are required to adapt this example to the general case. But (13.162) now implies that

$$\mathcal{A} \circ vv^\dagger = \lambda vv^\dagger \tag{13.166}$$

so that we have constructed an eigenmatrix of $\mathcal{A}$ with eigenvalue λ.

---

[11]To see this, one easily verifies that $\operatorname{tr}(\mathcal{B}) = 0 = \sigma(\mathcal{B})$, where $\mathcal{B} = \mathcal{A} - \sum \lambda_i \mathcal{P}_{\lambda_i}$. But this implies that $\operatorname{tr}(\mathcal{B}^2) = 0$, which forces $\mathcal{B} = 0$.

We can now perturb $\mathcal{A}$ slightly by adding $\epsilon\, vv^\dagger$, thus changing the eigenvalue of $vv^\dagger$ by ϵ. The resulting matrix will have three unequal eigenvalues, and hence admit a decomposition (13.161) in terms of orthogonal primitive idempotents. But these idempotents will also be eigenmatrices of $\mathcal{A}$, and hence yield an orthogonal primitive idempotent decomposition of $\mathcal{A}$.[12] In summary, decompositions analogous to (13.161) can also be found when there is a repeated eigenvalue, but the terms corresponding to the repeated eigenvalue cannot be written in terms of the projections $\mathcal{P}_\lambda$, and of course the decomposition of the corresponding eigenspace is not unique.[13]

13.7   Diagonalizing Jordan Matrices with F_4

The group F_4 is the automorphism group of both the Jordan and Freudenthal products, that is, if $\mathcal{X}, \mathcal{Y} \in \mathbf{H}_3(\mathbb{O})$, and $\phi \in F_4$, then

$$\phi(\mathcal{X} \circ \mathcal{Y}) = \phi(\mathcal{X}) \circ \phi(\mathcal{Y}), \qquad (13.171)$$

$$\phi(\mathcal{X} * \mathcal{Y}) = \phi(\mathcal{X}) * \phi(\mathcal{Y}). \qquad (13.172)$$

Loosely speaking, F_4 allows us to change basis in $\mathbf{H}_3(\mathbb{O})$, without affecting normalization or inner products. We have seen in Section 13.6 that there is indeed an eigenvalue problem for any Jordan matrix $\mathcal{X} \in \mathbf{H}_3(\mathbb{O})$, with exactly three real eigenvalues (counting multiplicity). We therefore expect

---

[12]More formally, with the above assumptions we have

$$(\mathcal{A} + \epsilon\, vv^\dagger - \lambda\mathcal{I}) * (\mathcal{A} + \epsilon\, vv^\dagger - \lambda\mathcal{I}) = (ww^\dagger + \epsilon\, vv^\dagger) * (ww^\dagger + \epsilon\, vv^\dagger)$$

$$= 2\epsilon\, vv^\dagger * ww^\dagger. \qquad (13.167)$$

The Freudenthal square of (13.167) is zero by (14.110), which shows that

$$\det(\mathcal{A} + \epsilon\, vv^\dagger - \lambda\mathcal{I}) = 0 \qquad (13.168)$$

by (13.146), so that λ is indeed an eigenvalue of the perturbed matrix $\mathcal{A} + \epsilon\, vv^\dagger$. Furthermore, (13.167) itself is not zero (unless v or w vanishes) since (13.165) implies that

$$2\,\mathrm{tr}\,(vv^\dagger * ww^\dagger) = (v^\dagger v)(w^\dagger w) \neq 0 \qquad (13.169)$$

which shows that λ does not have multiplicity two.

[13]An invariant orthogonal idempotent decomposition when λ is an eigenvalue of multiplicity two is

$$\mathcal{A} = \mu\,\frac{(\mathcal{A} - \lambda\mathcal{I})}{\mathrm{tr}\,(\mathcal{A} - \lambda\mathcal{I})} - \lambda\,\frac{(\widetilde{\mathcal{A} - \lambda\mathcal{I}})}{\mathrm{tr}\,(\mathcal{A} - \lambda\mathcal{I})} \qquad (13.170)$$

where the coefficient of $\mu = \mathrm{tr}\,(\mathcal{A}) - 2\lambda$ is the primitive idempotent corresponding to the other eigenvalue and the coefficient of λ is an idempotent but not primitive. An equivalent expression was given in [30].

to be able to find a basis in which $\mathcal{X}$ is diagonal. We show here that this expectation is correct, and any Jordan matrix can be diagonalized using F_4 transformations.

We start with a Jordan matrix in the form (13.77), and show how to diagonalize it using nested F_4 transformations. As discussed in [13], a set of generators for F_4 can be obtained by considering its SO(9) subgroups, which in turn can be generated by 2×2 tracefree, Hermitian, octonionic matrices.

Just as for the traditional diagonalization procedure, it is first necessary to solve the characteristic equation for the eigenvalues. Let λ be a solution of (13.75), and let $vv^\dagger \neq 0$ be a solution of (13.137) with eigenvalue λ.[14] We assume further that the phase in v is chosen such that

$$ v = \begin{pmatrix} x \\ y \\ r \end{pmatrix} \tag{13.173} $$

where $x, y \in \mathbb{O}$ and $r \in \mathbb{R}$. Define

$$ \mathcal{M}_1 = \frac{1}{N_1} \begin{pmatrix} -r & 0 & x \\ 0 & N_1 & 0 \\ \overline{x} & 0 & r \end{pmatrix}, \tag{13.174} $$

$$ \mathcal{M}_2 = \frac{1}{N_2} \begin{pmatrix} N_2 & 0 & 0 \\ 0 & -N_1 & y \\ 0 & \overline{y} & N_1 \end{pmatrix}, \tag{13.175} $$

where the normalization constants are given by $N_1^2 = |x|^2 + r^2$ and $N_2^2 = N_1^2 + |y|^2 = v^\dagger v \neq 0$. (If $N_1 = 0$, then $\mathcal{A}$ is already block diagonal.) It is straightforward to check that

$$ \mathcal{M}_2 \mathcal{M}_1 v = \begin{pmatrix} 0 \\ 0 \\ 1 \end{pmatrix} \tag{13.176} $$

and, since everything so far is quaternionic, this implies

$$ \mathcal{M}_2 \mathcal{M}_1 vv^\dagger \mathcal{M}_1 \mathcal{M}_2 = \begin{pmatrix} 0 & 0 & 0 \\ 0 & 0 & 0 \\ 0 & 0 & 1 \end{pmatrix} = \mathcal{E}_3 \tag{13.177} $$

since there are no associativity problems.

---

[14]It is straightforward to construct v using the results of Section 13.6, especially since we can assume without loss of generality that λ is an eigenvalue of multiplicity one.

But conjugation by each of the $\mathcal{M}_i$ is an F_4 transformation (which is well-defined since each $\mathcal{M}_i$ separately has components which lie in a *complex* subalgebra of $\mathbb{O}$); this is precisely the form of the generators referred to earlier. Furthermore, F_4 is the automorphism group of the Jordan product (13.138). Thus, since

$$(\mathcal{A} - \lambda\, vv^\dagger) \circ vv^\dagger = 0 \tag{13.178}$$

then after applying the (nested!) F_4 transformation above, we obtain

$$\big(\mathcal{M}_2(\mathcal{M}_1(\mathcal{A} - \lambda\mathcal{I})\mathcal{M}_1)\mathcal{M}_2\big) \circ \mathcal{E}_3 = 0 \tag{13.179}$$

which in turn forces

$$\mathcal{M}_2(\mathcal{M}_1\mathcal{A}\mathcal{M}_1)\mathcal{M}_2 = \begin{pmatrix} X & 0 \\ 0 & \lambda \end{pmatrix} \tag{13.180}$$

where

$$X = \begin{pmatrix} s & z \\ \overline{z} & t \end{pmatrix} \tag{13.181}$$

is a 2×2 octonionic Hermitian matrix (with $z \in \mathbb{O}$ and $s, t \in \mathbb{R}$).

The final step amounts to the diagonalization of X, which is easy. Let μ be any eigenvalue of X (which in fact means that it is another solution of (13.75)) and set

$$\mathcal{M}_3 = \frac{1}{N_3} \begin{pmatrix} \mu - t & 0 & 0 \\ 0 & t - \mu & z \\ 0 & \overline{z} & N_3 \end{pmatrix} \tag{13.182}$$

where $N_3 = (\mu - t)^2 + |z|^2$. (If $N_3 = 0$, X is already diagonal.) This finally results in

$$\mathcal{M}_3\big(\mathcal{M}_2(\mathcal{M}_1\mathcal{A}\mathcal{M}_1)\mathcal{M}_2\big)\mathcal{M}_3 = \begin{pmatrix} \mu & 0 & 0 \\ 0 & \mathrm{tr}\,(X) - \mu & 0 \\ 0 & 0 & \lambda \end{pmatrix} \tag{13.183}$$

and we have succeeded in diagonalizing $\mathcal{A}$ using F_4 as claimed.

Chapter 14

The Physics of the Octonions

14.1 Spin

We begin by considering angular momentum in quantum mechanics. Angular momentum is defined classically as the cross product of position and momentum, that is

$$\vec{L} = \vec{r} \times \vec{p}. \tag{14.1}$$

In quantum mechanics, momentum is replaced by a differential operator, that is

$$\vec{p} \longmapsto -i\hbar \vec{\nabla}. \tag{14.2}$$

Inserting (14.2) into (14.1) we obtain the components of the angular momentum operator as

$$L_x = -i\hbar(y\partial_z - z\partial_y), \tag{14.3}$$

$$L_y = -i\hbar(z\partial_x - x\partial_z), \tag{14.4}$$

$$L_z = -i\hbar(x\partial_y - y\partial_x). \tag{14.5}$$

These operators have the following characteristic properties. First of all, the commutator of any two is proportional to the third, so that

$$[L_x, L_y] = L_x L_y - L_y L_x = i\hbar L_z \tag{14.6}$$

and similarly for cyclic permutations. The second property is more subtle. The total angular momentum operator is

$$L^2 = |\vec{L}|^2 = \vec{L} \cdot \vec{L}. \tag{14.7}$$

When one separates variables in the Schrödinger equation in a central potential, the differential equation corresponding to the spherical angle θ reduces to an eigenvalue equation for the operator L^2. Imposing the boundary condition that the solutions should be well-behaved at the poles, the eigenvalues take the form $l(l + 1)\hbar^2$, where l is a non-negative integer.

The famous Stern–Gerlach experiment split a beam of atoms into separate beams for each of the $2l + 1$ allowed angular momentum states. If l is an integer, there should therefore be an odd number of beams. But the experiment produced two beams of silver atoms, thus showing that the angular momentum of the valence electron corresponds to $l = \frac{1}{2}$. How can we describe this phenomenon?

The commutation relations (14.6) (and cyclic permutations) can also be represented using matrices. A particularly nice choice is given in terms of the *Pauli matrices*

$$\boldsymbol{\sigma}_x = \begin{pmatrix} 0 & 1 \\ 1 & 0 \end{pmatrix}, \qquad \boldsymbol{\sigma}_y = \begin{pmatrix} 0 & -i \\ i & 0 \end{pmatrix}, \qquad \boldsymbol{\sigma}_z = \begin{pmatrix} 1 & 0 \\ 0 & -1 \end{pmatrix}, \qquad (14.8)$$

that were introduced in Section 7.3. If we now define

$$\boldsymbol{L}_a = \frac{\hbar}{2} \boldsymbol{\sigma}_a \qquad (14.9)$$

where $a = x, y, z$, then

$$[\boldsymbol{L}_x, \boldsymbol{L}_y] = \boldsymbol{L}_x \boldsymbol{L}_y - \boldsymbol{L}_y \boldsymbol{L}_x = i\hbar \boldsymbol{L}_z. \qquad (14.10)$$

If we interpret the $\boldsymbol{L}_m$ as the components of some sort of angular momentum, and compute the corresponding total angular momentum

$$\boldsymbol{L}^2 = \boldsymbol{L}_x^2 + \boldsymbol{L}_y^2 + \boldsymbol{L}_z^2 \qquad (14.11)$$

we discover that

$$\boldsymbol{L}^2 = \frac{3}{4} \hbar^2 \boldsymbol{I} \qquad (14.12)$$

since each of the Pauli matrices squares to the identity matrix. This description corresponds to $l = \frac{1}{2}$ (since $\frac{3}{4} = \frac{1}{2}(\frac{1}{2} + 1)$).

This new concept of angular momentum is however distinct from orbital angular momentum, and is therefore called spin angular momentum, or simply *spin*.[1]

Spinors represent states of spin-$\frac{1}{2}$ particles such as the electron. Spinors are important because they are needed to give a *spin-$\frac{1}{2}$ representation* of the angular momentum algebra (14.6).

----

[1]There is a well-known 3×3 representation of (14.6), involving matrices of the form

$$\hbar \begin{pmatrix} 0 & -i & 0 \\ i & 0 & 0 \\ 0 & 0 & 0 \end{pmatrix}$$

and cyclic permutations. The total angular momentum operator in this case—the sum of the squares of the three matrices—is $2\hbar$ times the identity, corresponding to $l = 1$; this is a *spin-1* representation of (14.6), corresponding to ordinary vectors in $\mathbb{R}^3$.

Here's a crash course in quantum mechanics. Physical states are represented by elements in a *Hilbert space*, which in this case consists of 2-component spinors. Any multiple of a given state v represents the same physical state, so we usually consider only normalized states, that is, we demand that $v^\dagger v = 1$. Operators such as $\boldsymbol{L}_z$ act on the state v by ordinary matrix multiplication, which takes v to some other state, given in this case by $\boldsymbol{L}_z v$. The *expectation value* of the operator $\boldsymbol{L}_z$ in the state v is given by $v^\dagger \boldsymbol{L}_z v$,[2] and gives the (average) result of (a large number of) physical measurements, in this case of the z-component of spin.

There are special states which satisfy the eigenvalue equation

$$\boldsymbol{L}_z v_\lambda = \lambda v_\lambda \qquad (14.13)$$

for some λ. The *eigenvalues* λ are the only possible results of physical measurements; it is only in an *eigenstate* v_λ that the measured value actually equals the expectation value. By requiring physical quantities ("observables") to correspond to Hermitian matrices, we ensure that the result of any measurement is real, since the eigenvalues of Hermitian matrices are real—at least, over $\mathbb{C}$. (See Chapter 13 for a discussion of what happens over $\mathbb{H}$ and $\mathbb{O}$.)

What are the eigenstates of $\boldsymbol{L}_z$? We have

$$\boldsymbol{L}_z v_\pm = \pm \frac{\hbar}{2} v_\pm \qquad (14.14)$$

where

$$v_+ = \begin{pmatrix} 1 \\ 0 \end{pmatrix}, \qquad v_- = \begin{pmatrix} 0 \\ 1 \end{pmatrix}. \qquad (14.15)$$

The expectation value of either $\boldsymbol{L}_x$ or $\boldsymbol{L}_y$ is 0 in these states: A "spin-up" electron is equally likely to be "spin-left" as "spin-right". Furthermore, $v_\pm$ are *not* eigenstates of either $\boldsymbol{L}_x$ or $\boldsymbol{L}_y$. Only operators which commute with each other correspond to observables that can be measured simultaneously. It is therefore not possible to simultaneously measure more than one component of angular momentum!

As discussed in Section 7.3, the Pauli matrices are also closely related to rotations in three dimensions—as they should be if they are to describe

---

[2] Those familiar with the Dirac "bra/ket" notation can make the identifications

$$|v\rangle \longleftrightarrow v,$$
$$\langle v| \longleftrightarrow v^\dagger.$$

angular momentum. Using (14.9), we see that

$$L_z = i\hbar \frac{dR_z}{d\alpha}\bigg|_{\alpha=0} \tag{14.16}$$

and similarly for L_y and L_x. The spin operators are thus infinitesimal rotations; they live in the Lie algebra corresponding to the Lie group of 3-dimensional rotations.

14.2   Quaternionic Spin

In quantum mechanics, the eigenstates of a self-adjoint operator correspond to the physical states with particular values of the corresponding observable physical quantity. A fundamental principle of quantum mechanics further states that after making a measurement the system is "projected" into the corresponding eigenstate. It is therefore only eigenstates which are unaffected by the measurement process; they are projected into themselves. In particular, in order to simultaneously make *two* measurements, the system must be in an eigenstate of each; otherwise, the order in which the measurements are made will matter.

As discussed in Section 14.1, the spin operators L_m are imaginary multiples of the derivatives ("infinitesimal generators") of the rotations R_m introduced originally in Section 7.3. Writing r_a for those derivatives, we have

$$r_a = \frac{dR_z}{d\alpha}\bigg|_{\alpha=0} \tag{14.17}$$

so that

$$r_x = \frac{1}{2}\begin{pmatrix} 0 & \ell \\ \ell & 0 \end{pmatrix}, \qquad r_y = \frac{1}{2}\begin{pmatrix} 0 & 1 \\ -1 & 0 \end{pmatrix}, \qquad r_z = \frac{1}{2}\begin{pmatrix} \ell & 0 \\ 0 & -\ell \end{pmatrix}, \tag{14.18}$$

where we have used ℓ rather than i to denote the complex unit. One then normally multiplies by $-\ell$ (and adds a factor of $\hbar$, which we henceforth set to 1) to obtain a description of the Lie algebra $\mathfrak{su}(2)$ in terms of Hermitian matrices.

As discussed in [34, 35], however, even in the quaternionic setting care must be taken with this last step; we must put the factor of ℓ in the right place! The right place turns out to be on the right; we define the operator

$$\widehat{L}_m(\psi) = -(r_m\psi)\ell \tag{14.19}$$

where ψ is a 2-component *octonionic* column (representing a Majorana-Weyl spinor in ten spacetime dimensions). The operators $\widehat{\boldsymbol{L}}_m$ are self-adjoint with respect to the inner product

$$\langle \psi, \chi \rangle = \pi\left(\psi^\dagger \chi\right) \tag{14.20}$$

where the map

$$\pi(q) = \frac{1}{2}(q + \ell q \bar{\ell}) \tag{14.21}$$

projects $\mathbb{O}$ to a preferred complex subalgebra $\mathbb{C} \subset \mathbb{O}$, namely the one containing ℓ.

Spin eigenstates are the eigenvectors of L_z, whose eigenvalues are $\pm\frac{1}{2}$ (really $\pm\frac{\hbar}{2}$). What are the eigenvectors of $\widehat{\boldsymbol{L}}_z$? Unsurprisingly, we have

$$\widehat{\boldsymbol{L}}_z \begin{pmatrix} 1 \\ 0 \end{pmatrix} = \begin{pmatrix} 1 \\ 0 \end{pmatrix} \frac{1}{2}. \tag{14.22}$$

(We could of course have written the eigenvalue on the left since it is real.) Somewhat surprisingly, this is not the only eigenvector with eigenvalue $\frac{1}{2}$. For instance, we have

$$\widehat{\boldsymbol{L}}_z \begin{pmatrix} 0 \\ k \end{pmatrix} = \begin{pmatrix} 0 \\ k \end{pmatrix} \frac{1}{2}. \tag{14.23}$$

Note the crucial role played by the anticommutativity of the quaternions in this equation! Particular attention is paid in [34, 35] to the eigenstates

$$e_\uparrow = \begin{pmatrix} 1 \\ k \end{pmatrix}, \qquad e_\downarrow = \begin{pmatrix} -k \\ 1 \end{pmatrix}, \tag{14.24}$$

which satisfy

$$\widehat{\boldsymbol{L}}_z(e_\uparrow) = \frac{1}{2}e_\uparrow, \qquad \widehat{\boldsymbol{L}}_z(e_\downarrow) = -\frac{1}{2}e_\downarrow, \tag{14.25}$$

and which were proposed as representing particles[3] at rest with spin $\pm\frac{1}{2}$, respectively. The eigenstates $e_\uparrow$ and $e_\downarrow$ are orthogonal with respect to the above inner product, that is

$$\langle e_\uparrow, e_\downarrow \rangle = 0. \tag{14.26}$$

We will therefore focus on these eigenstates, which have some extraordinary properties.

---

[3] These two papers used different conventions to distinguish particles from antiparticles; we adopt the conventions used in [35].

Consider now the remaining spin operators $\widehat{L}_x$, $\widehat{L}_y$ acting on these eigenstates. We have

$$\widehat{L}_x(e_\uparrow) = \frac{1}{2}\begin{pmatrix} -k \\ 1 \end{pmatrix} = e_\uparrow\left(-\frac{k}{2}\right) \qquad (14.27)$$

and

$$\widehat{L}_y(e_\uparrow) = \frac{1}{2}\begin{pmatrix} -k\ell \\ \ell \end{pmatrix} = e_\uparrow\left(-\frac{k\ell}{2}\right) \qquad (14.28)$$

with similar results holding for $e_\downarrow$, illustrating the fact that this *quaternionic* self-adjoint operator eigenvalue problem admits eigenvalues which are not real. More importantly, as claimed in [34, 35], it shows that $e_\uparrow$ is a *simultaneous* eigenvector of the three self-adjoint spin operators $\widehat{L}_x$, $\widehat{L}_y$, and $\widehat{L}_z$!

This result could have significant implications for quantum mechanics. In this formulation, the inability to completely measure the spin state of a particle, because the spin operators fail to commute, is thus ultimately due to the fact that the *eigenvalues* don't commute. Explicitly, we have

$$4\widehat{L}_x\big(\widehat{L}_y(e_\uparrow)\big) = 2\widehat{L}_x(-e_\uparrow\,k\ell) = 2r_x(e_\uparrow\,k\ell)\,\ell$$
$$= -2r_x(e_\uparrow\,\ell)\,k\ell = +2\widehat{L}_x(e_\uparrow)\,k\ell$$
$$= -e_\uparrow\,k\,k\ell = +e_\uparrow\,\ell, \qquad (14.29)$$

$$4\widehat{L}_y\big(\widehat{L}_x(e_\uparrow)\big) = -e_\uparrow\,k\ell\,k = -e_\uparrow\,\ell, \qquad (14.30)$$

which yields the usual commutation relation in the form

$$[\widehat{L}_x, \widehat{L}_y]\,(e_\uparrow) = \frac{1}{2}\,e_\uparrow\,\ell = \widehat{L}_z(e_\uparrow)\,\ell. \qquad (14.31)$$

Furthermore, there is a phase freedom in (14.24), since

$$\widehat{L}_z\left(e_\uparrow e^{\ell\theta}\right) = \left(e_\uparrow e^{\ell\theta}\right)\left(\frac{1}{2}\right) \qquad (14.32)$$

for any value of θ. It is still true that $e_\uparrow e^{\ell\theta}$ is a simultaneous eigenvector of all three spin operators, but the imaginary eigenvalues have changed. We have

$$\widehat{L}_x\left(e_\uparrow e^{\ell\theta}\right) = \left(e_\uparrow e^{\ell\theta}\right)\left(-\frac{k\,e^{2\ell\theta}}{2}\right), \qquad (14.33)$$

$$\widehat{L}_y\left(e_\uparrow e^{\ell\theta}\right) = \left(e_\uparrow e^{\ell\theta}\right)\left(-\frac{k\ell\,e^{2\ell\theta}}{2}\right), \qquad (14.34)$$

so that the non-real eigenvalues depend on the phase. It is intriguing to speculate on whether the value of the non-real eigenvalues, which determine the phase, can be used to specify (but not measure) the actual direction of the spin, and whether this might shed some insight on basic properties of quantum mechanics such as Bell's inequality.

Finally, we point out that all eigenvectors of the complex operators L_x, L_y, L_z turn out to be quaternionic; each eigenvector lies in some quaternionic subalgebra of $\mathbb{O}$ which also contains ℓ.

14.3 Introduction to the Dirac Equation

The *Dirac equation* describes the quantum mechanical state of a relativistic, massive, spin-$\frac{1}{2}$ particle, such as the electron. The Dirac equation in four dimensions is usually given as

$$(i\hbar\gamma^\mu\partial_\mu - mc)\Psi = 0. \tag{14.35}$$

What do these symbols mean? First of all, m is the mass of the particle described by Ψ, c is the speed of light (which we normally set to 1) and $\hbar$ is Planck's constant (divided by 2π). The notation ∂_μ is short for $\frac{\partial}{\partial x^\mu}$, and the repeated index μ implies a summation over the spacetime index $\mu = 0, 1, 2, 3$, where $x^0 = ct$, $x^1 = x$, $x^2 = y$, and $x^3 = z$. But what is γ^μ? We will briefly postpone this question.

The form (14.35) assumes that the "squared length" of the spacetime vector whose components are x^μ is given by[4]

$$g_{\mu\nu}x^\mu x^\nu = (x^0)^2 - (x^1)^2 - (x^2)^2 - (x^3)^2 \tag{14.36}$$

so that the *spacetime metric* is given by the matrix

$$(g_{\mu\nu}) = \begin{pmatrix} 1 & 0 & 0 & 0 \\ 0 & -1 & 0 & 0 \\ 0 & 0 & -1 & 0 \\ 0 & 0 & 0 & -1 \end{pmatrix}. \tag{14.37}$$

What happens if we multiply the Dirac equation (14.35) by the differential operator obtained by replacing m with $-m$? We get

$$0 = (i\hbar\gamma^\mu\partial_\mu + mc)(i\hbar\gamma^\nu\partial_\nu - mc)\Psi = -(\hbar^2\gamma^\mu\partial_\mu\gamma^\nu\partial_\nu + m^2c^2)\Psi \tag{14.38}$$

---

[4]These signs determine a choice of *signature*. The other choice is to use $-g_{\mu\nu}$ for the metric components, which eliminates the i in (14.35) (and multiplies all the γ matrices by i).

where we have been careful to use different dummy indices in the two implicit sums. The right-hand side almost looks like the Klein–Gordon equation,

$$(\hbar^2\Box + m^2c^2)\phi = 0, \tag{14.39}$$

a wave equation which describes the state of a relativistic, massive, spin-0 particle. Here, $\Box$ is the *d'Alembertian* operator, the spacetime version of the Laplacian, which is given by

$$\Box = g^{\mu\nu}\partial_\mu\partial_\nu = \partial_t^2 - \triangle = \partial_t^2 - \nabla^2 \tag{14.40}$$

where $(g^{\mu\nu})$ is the inverse metric, that is, the inverse of the matrix $(g_{\mu\nu})$.

The Klein–Gordon equation was known to Dirac, but there were difficulties (later resolved) interpreting the squares of its solutions as probability densities, arising from the equation being second order. Dirac was led to his first-order equation by reversing the argument just given, thus "factoring" the Klein–Gordon equation. To make this work, he had to show that (14.38) really is the Klein–Gordon equation.

Assume that each γ^μ is constant. Then we must solve the equation

$$\gamma^\mu\gamma^\nu\partial_\mu\partial_\nu = g^{\mu\nu}\partial_\mu\partial_\nu. \tag{14.41}$$

Writing out all the terms of (14.41) and comparing coefficients of $\partial_\mu\partial_\nu$, remembering that partial derivative operators commute ($\partial_\mu\partial_\nu = \partial_\nu\partial_\mu$), we obtain

$$\{\gamma^\mu, \gamma^\nu\} = \gamma^\mu\gamma^\nu + \gamma^\nu\gamma^\mu = g^{\mu\nu} + g^{\nu\mu} = 2g^{\mu\nu} \tag{14.42}$$

where we have introduced the curly bracket notation for *anticommutators*. It is easily seen that (14.42) has no solutions if each γ^μ is a number. Dirac's brilliant idea was to use matrices instead, which we will discuss in Section 14.4.

Suppose that Ψ is a plane-wave of the form

$$\Psi = e^{-ip_\nu x^\nu/\hbar}\Psi_0 \tag{14.43}$$

where p_ν is the 4-momentum of the particle and where Ψ_0 does not depend on x^μ. Inserting (14.43) into (14.35) leads to the *momentum space* Dirac equation

$$(\gamma^\mu p_\mu - mc)\Psi_0 = 0 \tag{14.44}$$

which is purely algebraic.[5] (We will usually write Ψ rather than Ψ_0.)

---

[5]One can also view (14.44) as the Fourier transform of (14.35). Equivalently, one obtains (14.44) from (14.35) by the formal substitution $i\hbar\partial_\mu \mapsto p_\mu$.

Note the absence of any explicit factors of i in the momentum-space Dirac equation (14.44)! We view momentum space as more fundamental than position space, and deliberately chose the metric signature to achieve this property. It is now straightforward to generalize (14.44) to higher dimensions using the other division algebras, which would not have been the case had there been factors of i to worry about.

14.4 Gamma Matrices

In Section 14.3, we introduced the momentum-space Dirac equation, which we rewrite in the form[6]

$$(\gamma_\mu \, p^\mu - m)\Psi = 0 \qquad (14.45)$$

where we have set $c = 1$. The gamma matrices γ_μ satisfy

$$\{\gamma_\mu, \gamma_\nu\} = \gamma_\mu \gamma_\nu + \gamma_\nu \gamma_\mu = 2g_{\mu\nu}. \qquad (14.46)$$

How do we find such matrices?

It turns out that, in all the cases we will be interested in, the gamma matrices can be constructed in blocks from the Pauli matrices $\boldsymbol{\sigma}_a$. We begin by discussing how to construct big matrices from small ones in this way.

Imagine a chessboard. It's an 8×8 grid. Now suppose it's a 3-dimensional chess game. Easy; just stack eight boards on top of each other. But there's no need to stack them vertically! Simply put eight ordinary chess boards next to each other, and you can still play 3-dimensional chess. What about four dimensions? Take eight rows of eight chessboards each. You can keep going to get a chess game in any (finite!) dimension, but four is enough for our needs.

How do you label a square in our 4-dimensional chess game? You need to specify both the chessboard, and the square on the chessboard. But each of these specifications corresponds to an element of an 8×8 matrix! Thus, a square on the chessboard is labeled by specifying an element of *two* 8×8 matrices.

Of course, we can also imagine this setup as a single 64×64 board, whose squares are specified by giving an element of a 64×64 matrix. We have thus built up a 64×64 matrix using two 8×8 matrices, one to describe

---

[6]Indices can be raised and lowered using the metric; a time component (index 0) is unchanged, while a space component (indices 1–3) picks up a minus sign. A pair of indices being summed over should always have one up and one down, in which case it doesn't matter which is which.

the arrangement of the blocks, the other to describe the location in each block.

Consider doing this construction with 2×2 matrices, rather than 8×8. Suppose the first is

$$\sigma_1 = \sigma_x = \begin{pmatrix} 0 & 1 \\ 1 & 0 \end{pmatrix} \tag{14.47}$$

and the second is the identity matrix. What is the result?

The first matrix, σ_1, gives the block structure: The upper-left and lower-right blocks are (multiplied by) 0, while the remaining blocks are (multiplied by) 1. What's in each block? The identity matrix! What is the result? We write

$$\sigma_1 \otimes I = \begin{pmatrix} 0 & 0 & 1 & 0 \\ 0 & 0 & 0 & 1 \\ 1 & 0 & 0 & 0 \\ 0 & 1 & 0 & 0 \end{pmatrix} \tag{14.48}$$

where the symbol $\otimes$ is read "tensor"; this is a *tensor product*. Tensors are generalizations of matrices, in this case a 4-dimensional array of numbers, which is $2 \times 2 \times 2 \times 2$. Just as with the chessboards, we reinterpret this as an ordinary matrix, which in this case is 4×4.

The power of this description comes from the fact that matrix multiplication is compatible with the tensor product, in the sense that

$$(A_1 \otimes B_1)(A_2 \otimes B_2) = (A_1 A_2) \otimes (B_1 B_2). \tag{14.49}$$

Compatibility doesn't quite work for anticommutators, except in special cases. One such case occurs if either $[A_1, A_2] = 0$ or $[B_1, B_2] = 0$, for which

$$\{(A_1 \otimes B_1), (A_2 \otimes B_2)\} = \frac{1}{2}\{A_1, A_2\} \otimes \{B_1, B_2\}. \tag{14.50}$$

By constructing the gamma matrices as tensor products of Pauli matrices, we can use (14.49) and (14.50) to work out products of gamma matrices in terms of products of Pauli matrices, which are much easier. Remembering that

$$\sigma_2 = \sigma_y = \begin{pmatrix} 0 & -i \\ i & 0 \end{pmatrix}, \qquad \sigma_3 = \sigma_z = \begin{pmatrix} 1 & 0 \\ 0 & -1 \end{pmatrix}, \tag{14.51}$$

it is easy to check that the Pauli matrices anticommute and square to the identity, that is,

$$\{\sigma_a, \sigma_b\} = 2\delta_{ab} \tag{14.52}$$

where δ_{ab} is the Kronecker delta, which is 1 if $a = b$ and 0 otherwise.

There are many possible choices of matrices γ_μ which satisfy (14.46); each such choice is called a *representation* (more formally, a representation of the underlying Clifford algebra). We seek a *minimal* representation, that is, we are looking for the smallest matrices possible, so that the solutions Ψ of the Dirac equation have as few physical degrees of freedom as possible. In four dimensions, the smallest matrices that can be used are 4×4, but there are still many representations.

We choose a representation which will generalize nicely to the other division algebras, namely

$$\gamma_0 = \boldsymbol{\sigma}_1 \otimes \boldsymbol{I} = \begin{pmatrix} 0 & \boldsymbol{I} \\ \boldsymbol{I} & 0 \end{pmatrix}, \tag{14.53}$$

$$\gamma_a = i\boldsymbol{\sigma}_2 \otimes \boldsymbol{\sigma}_a = \begin{pmatrix} 0 & \boldsymbol{\sigma}_a \\ -\boldsymbol{\sigma}_a & 0 \end{pmatrix}, \tag{14.54}$$

where $a = 1, 2, 3$. These expressions are carefully chosen so that, given any two distinct gamma matrices, exactly one of the factors commutes. Using (14.50), it is then obvious that distinct gamma matrices anticommute; all that remains to be checked is that they square to the correct multiples of the identity matrix. They do.

This representation is not the one found in most introductory textbooks. It has the advantage that all of the gamma matrices have the same off-diagonal block structure. This structure emphasizes the fundamental role played by the mass term in the Dirac equation, which multiplies the identity matrix. Such a representation is called a *chiral* or *Weyl* representation (because the eigenspinors of the chiral projection operator take on a particularly nice form). Multiplying the Dirac equation (14.45) on the left by γ_0 brings it to a form equivalent to Dirac's original formulation, namely

$$(\gamma_0\gamma_\mu\, p^\mu - m\gamma_0)\Psi = 0. \tag{14.55}$$

This form emphasizes the chiral nature of the representation: The first term is block diagonal, and the mass term "couples" the otherwise unrelated blocks. Furthermore, both matrix coefficients ($\gamma_0\gamma_\mu$ and γ_0) are now Hermitian—and both square to the identity matrix.

We have not yet said what Ψ is. Since the gamma matrices are matrices, Ψ is clearly a column vector of the appropriate size. However, it turns out that Ψ transforms in a particular way under the Lorentz group; Ψ is a (Dirac) *spinor*; the description "column vector" is misleading, and will be avoided.

The set of all products of gamma matrices is the basic example of a *Clifford algebra*. Using the anticommutativity properties, any such product can be simplified so that it contains each gamma matrix at most once. Each element of the Clifford algebra can therefore be classified as even or odd depending on the number of gamma matrices it contains. The product of an even element with another even element is still even; the even part of the Clifford algebra is a subalgebra. In the chiral representation above, the even elements are block diagonal, and the odd elements are block off-diagonal.

Our gamma matrices have the further advantage that they generalize immediately to the other division algebras. First of all, the apparent factor of i in (the first factor of) γ_a isn't really there; $i\sigma_2$ is real. We can expand our set of four Pauli matrices by including (in the second factor of γ_a) the generalized Pauli matrices obtained from σ_2 by replacing i by j, k, etc.[7] There are as many of these matrices as there are imaginary units in the division algebra; they still anticommute, by virtue of the division algebra multiplication table, so that (14.50) still holds. Remembering to include σ_x and σ_z, we get 3, 4, 6, or 10 gamma matrices over $\mathbb{R}$, $\mathbb{C}$, $\mathbb{H}$, and $\mathbb{O}$, respectively. We will use this construction in Section 14.5 to discuss the Dirac equation in higher dimensions.

14.5   The Dirac Equation

It's time to put everything together. Insert (14.53) and (14.54) into (14.55). Take advantage of the block structure by writing the (4-component) Dirac spinor Ψ in terms of two (2-component) Penrose/Weyl spinors θ and η as

$$\Psi = \begin{pmatrix} \theta \\ \eta \end{pmatrix} \tag{14.56}$$

which yields a particularly nice form of the Dirac equation, namely

$$\begin{pmatrix} p^0 \boldsymbol{I} - p^a \boldsymbol{\sigma}_a & -m\boldsymbol{I} \\ -m\boldsymbol{I} & p^0 \boldsymbol{I} + p^a \boldsymbol{\sigma}_a \end{pmatrix} \begin{pmatrix} \theta \\ \eta \end{pmatrix} = 0. \tag{14.57}$$

We can do even better. The 2×2 matrix corresponding to the momentum vector p^μ is just

$$\boldsymbol{P} = p^\mu \boldsymbol{\sigma}_\mu \tag{14.58}$$

---

[7] We number the generalized Pauli matrices sequentially from 2, renumbering σ_3 to put it last. To avoid confusion, we will often use the indices x, y, z for the original Pauli matrices, and j, k, etc. for the generalized Pauli matrices.

and we introduce the notation

$$\widetilde{\boldsymbol{P}} = \boldsymbol{P} - \operatorname{tr}(\boldsymbol{P})I \qquad (14.59)$$

for trace reversal, which reverses the sign of p^0.[8] Working out (14.57), the Dirac equation reduces to the two equations

$$-\widetilde{\boldsymbol{P}}\theta - m\eta = 0, \qquad (14.60)$$
$$-m\theta + \boldsymbol{P}\eta = 0, \qquad (14.61)$$

which no longer contain any gamma matrices! Furthermore, inserting (14.61) into (14.60) yields the constraint

$$-\widetilde{\boldsymbol{P}}\boldsymbol{P} = m^2 I \qquad (14.62)$$

or equivalently

$$p_\mu p^\mu = m^2 \qquad (14.63)$$

which is just the condition that the norm of the momentum vector be the mass.

This construction works over any of the four division algebras, and over the complex numbers we recover the usual formalism in four spacetime dimensions. In this case, θ and η have two complex components. But since

$$\mathbb{H}^2 = \mathbb{C}^2 \oplus \mathbb{C}^2 \qquad (14.64)$$

we can replace a pair of complex 2-component spinors by a single quaternionic 2-component spinor. We choose the identification

$$\begin{pmatrix} A \\ B \\ C \\ D \end{pmatrix} \longleftrightarrow \begin{pmatrix} C - kB \\ D + kA \end{pmatrix} \qquad (14.65)$$

with $A, B, C, D \in \mathbb{C}$. Equivalently, we can write this identification in terms of the Penrose/Weyl spinors θ and η as

$$\Psi = \begin{pmatrix} \theta \\ \eta \end{pmatrix} \longleftrightarrow \psi = \eta + \boldsymbol{\sigma}_k \theta \qquad (14.66)$$

---

[8]We can therefore think of $-\widetilde{\boldsymbol{P}}$ as the 1-form (covariant vector) that is dual to $\boldsymbol{P}$. This interpretation is strengthened by noting that

$$-\widetilde{\boldsymbol{P}}\boldsymbol{P} = \det(\boldsymbol{P})I = p_\mu p^\mu I.$$

where the generalized Pauli matrix $\boldsymbol{\sigma}_k$ is given by

$$\boldsymbol{\sigma}_k = \begin{pmatrix} 0 & -k \\ k & 0 \end{pmatrix}. \tag{14.67}$$

Since (14.66) is clearly a vector space isomorphism, there is also an isomorphism relating the linear maps on these spaces. We can use the induced isomorphism to rewrite the (4-dimensional, complex) Dirac equation (14.55) in 2-component quaternionic language. Direct computation yields the correspondences

$$\gamma_0 \gamma_a \longleftrightarrow \boldsymbol{\sigma}_a \tag{14.68}$$

and

$$\gamma_0 \longleftrightarrow \boldsymbol{\sigma}_k. \tag{14.69}$$

One way to see that these equivalences make sense is to notice that

$$\Psi = \begin{pmatrix} \theta \\ 0 \end{pmatrix} + \gamma_0 \begin{pmatrix} \eta \\ 0 \end{pmatrix}. \tag{14.70}$$

Direct translation of (14.55) now leads to the quaternionic Dirac equation

$$(\boldsymbol{P} - m\boldsymbol{\sigma}_k)(\eta + \boldsymbol{\sigma}_k\theta) = 0. \tag{14.71}$$

Working backward, we can separate (14.71) into an equation not involving k, which is precisely (14.61), and an equation involving k, which is

$$\boldsymbol{P}\boldsymbol{\sigma}_k\theta - m\boldsymbol{\sigma}_k\eta = 0. \tag{14.72}$$

Multiplying (14.72) on the left by $\boldsymbol{\sigma}_k$, and using the remarkable identity

$$\boldsymbol{\sigma}_k\boldsymbol{P}\boldsymbol{\sigma}_k = -\widetilde{\boldsymbol{P}} \tag{14.73}$$

reduces (14.72) to (14.60), as expected.

So far, all we have done is rewrite the usual (4-dimensional, complex) Dirac equation in 2-component quaternionic language. However, the appearance of the term $m\boldsymbol{\sigma}_k$ suggests a way to put the mass term on the same footing as the other terms, which we now exploit. Multiplying (14.71) on the left by $-\boldsymbol{\sigma}_k$ and using (14.73) brings the (4-dimensional) Dirac equation to the form

$$\left(\widetilde{\boldsymbol{P}} + m\boldsymbol{\sigma}_k\right)\left(\theta + \boldsymbol{\sigma}_k\eta\right) = 0. \tag{14.74}$$

The spinor in parentheses is just $\boldsymbol{\sigma}_k \psi$, which we will henceforth relabel as simply ψ, that is, from now on we write[9]

$$\psi = \boldsymbol{\sigma}_k(\eta + \boldsymbol{\sigma}_k\theta) = \theta + \boldsymbol{\sigma}_k\eta. \tag{14.75}$$

When written out in full, (14.74) takes the form

$$\begin{pmatrix} -p^t + p^z & p^x - \ell p^y - km \\ p^x + \ell p^y + km & -p^t - p^z \end{pmatrix} \psi = 0. \tag{14.76}$$

which clearly suggests viewing the mass as an additional spacelike component of a higher-dimensional vector. Furthermore, since the matrix multiplying ψ has determinant zero, this higher-dimensional vector is null. We thus appear to have reduced the massive Dirac equation in four dimensions to the massless Dirac, or Weyl, equation in higher dimensions, thus putting the massive and massless cases on an equal footing. This expectation is indeed correct, as we will show below in the more general octonionic setting.

14.6 The Weyl Equation

Consider the Dirac equation (14.57) with $m = 0$. Then equations (14.60) and (14.61) decouple, so it is enough to consider just one, say (14.60). This is the *Weyl* equation

$$\widetilde{\boldsymbol{P}}\psi = 0 \tag{14.77}$$

where we have written ψ instead of θ. In matrix notation, it is straightforward to show that the momentum p^μ of a solution of the Weyl equation must be null: (14.77) says that the 2×2 Hermitian matrix $\boldsymbol{P}$ has 0 as one of its eigenvalues, which forces $\det(\boldsymbol{P}) = 0$, which is precisely the condition that p^μ be null.

Note that $\boldsymbol{P}$ is a *complex* matrix; it contains only one octonionic direction. But a 2×2 complex Hermitian matrix with determinant zero can be written as

$$\boldsymbol{P} = \pm\theta\theta^\dagger \tag{14.78}$$

where θ is also complex. The general solution of (14.77) is

$$\psi = \theta\xi \tag{14.79}$$

---

[9] This change in notation also inserts a minus sign into the correspondence (14.68), but leaves (14.69) intact. We could of course have simply started with (14.75) in (14.65), but that choice makes it slightly more difficult to obtain (14.71).

where $\xi \in \mathbb{O}$ is arbitrary. It follows immediately from (14.79) that

$$\psi \psi^\dagger = \pm |\xi|^2 \boldsymbol{P} \qquad (14.80)$$

which says that the vector constructed from ψ is proportional to $\boldsymbol{P}$.

Since there are still just two octonions in all, ξ and the components of θ (and hence also those of $\boldsymbol{P}$) belong to a *quaternionic* subalgebra of $\mathbb{O}$. Thus, for solutions (14.79), the Weyl equation (14.77) itself becomes quaternionic! We can assume without loss of generality that this quaternionic subalgebra is the one containing k and ℓ. We therefore have

$$\boldsymbol{P} = p^t \boldsymbol{I} + p^x \boldsymbol{\sigma}_x + p^y \boldsymbol{\sigma}_y + p^z \boldsymbol{\sigma}_z + p^k \boldsymbol{\sigma}_k + p^{k\ell} \boldsymbol{\sigma}_{k\ell}. \qquad (14.81)$$

We can further assume, by a rotation in the plane containing k and $k\ell$ if necessary, that $p^{k\ell} = 0$. Writing $m = p^k$ brings $\boldsymbol{P}$ precisely to the form (14.76)! The *octonionic* Weyl equation, describing massless spin-$\frac{1}{2}$ particles in ten dimensions, can therefore be reduced to the complex *Dirac* equation, describing, in general, massive spin-$\frac{1}{2}$ particles in four dimensions. We will pursue this program in the next section.

14.7   Leptons[10]

The description in the preceding sections of 10-dimensional Minkowski space in terms of Hermitian octonionic matrices is a direct generalization of the usual description of ordinary (4-dimensional) Minkowski space in terms of complex Hermitian matrices. If we fix a complex subalgebra $\mathbb{C} \subset \mathbb{O}$, then we single out a 4-dimensional Minkowski subspace of 10-dimensional Minkowski space. The projection of a 10-dimensional null vector onto this subspace is a causal 4-dimensional vector, which is null if and only if the original vector was already contained in the subspace, and timelike otherwise. The time orientation of the projected vector is the same as that of the original, and the induced mass is given by the norm of the remaining six components. Furthermore, the ordinary Lorentz group $\mathrm{SO}(3,1)$ clearly sits inside the Lorentz group $\mathrm{SO}(9,1)$ via the identification of their double covers, the spin groups $\mathrm{Spin}(d,1)$, namely

$$\mathrm{Spin}(3,1) = \mathrm{SL}(2,\mathbb{C}) \subset \mathrm{SL}(2,\mathbb{O}) = \mathrm{Spin}(9,1). \qquad (14.82)$$

Therefore, all it takes to break ten spacetime dimensions to four is to choose a preferred octonionic unit to play the role of the complex unit.

---

[10]The material in this section is adapted from [35].

We choose ℓ rather than i to fill this role, preferring to save i, j, k for a (distinguished) quaternionic triple. The projection π from $\mathbb{O}$ to $\mathbb{C}$ is then

$$\pi(q) = \frac{1}{2}(q + \ell q \bar{\ell}) \tag{14.83}$$

and we thus obtain a preferred SL(2, $\mathbb{C}$) subgroup of SL(2, $\mathbb{O}$), corresponding to the "physical" Lorentz group.

For each solution ψ of (14.77), the momentum is proportional to $\psi\psi^\dagger$ by (14.80). Up to an overall factor, we can therefore read off the components of the 4-dimensional momentum p^μ directly from $\pi(\psi\psi^\dagger)$. The projection of a 10-dimensional lightlike vector to four dimensions results in a causal vector, that is, in a vector which is either lightlike or timelike. In other words, the resulting 4-momentum is that of a massless or massive particle, respectively. We can use a (4-dimensional!) Lorentz transformation to bring a massive particle to rest, or to orient the momentum of a massless particle to be in the z-direction.

As in the previous chapter, we will assume that the components of Ψ— and hence also of $\boldsymbol{P}$—lie in the quaternionic subalgebra containing k and ℓ, and that $p^{k\ell} = 0$. We therefore have

$$\boldsymbol{P} = \pi(\boldsymbol{P}) + m\boldsymbol{\sigma}_k \tag{14.84}$$

which shows explicitly the relationship between the higher-dimensional lightlike vector $\boldsymbol{P}$, the causal 4-dimensional vector $\pi(\boldsymbol{P})$, and the mass m.

If $m \neq 0$, we can distinguish particles from antiparticles by the sign of the term involving m, which is the coefficient of $\boldsymbol{\sigma}_k$ in $\boldsymbol{P}$. Equivalently, we have the particle/antiparticle projections (at rest)

$$\boldsymbol{\Pi}_\pm = \frac{1}{2}\left(\boldsymbol{I} \pm \boldsymbol{\sigma}_k\right). \tag{14.85}$$

If $m = 0$, however, we can only distinguish particles from antiparticles in momentum space by the sign of p^t, as usual; this is the same as the sign in (14.80). Similarly, in this language, the chiral projection operator is constructed from

$$\boldsymbol{\Upsilon}_5 = \boldsymbol{\sigma}_t \boldsymbol{\sigma}_x \boldsymbol{\sigma}_y \boldsymbol{\sigma}_z = -\begin{pmatrix} \ell & 0 \\ 0 & \ell \end{pmatrix}. \tag{14.86}$$

However, as with spin, we must multiply by ℓ in the correct place, that is

$$\widehat{\boldsymbol{\Upsilon}}_5[\psi] = \boldsymbol{\Upsilon}_5 \psi \ell. \tag{14.87}$$

As a result, even though $\boldsymbol{\Upsilon}_5$ is a multiple of the identity, $\widehat{\boldsymbol{\Upsilon}}_5$ is not, and the operators $\frac{1}{2}(\boldsymbol{I} \pm \widehat{\boldsymbol{\Upsilon}}_5)$ project $\mathbb{H}^2$ into the Weyl subspaces $\mathbb{C}^2 \oplus \mathbb{C}^2 k$ as desired.

Combining the spin and particle information, over the quaternionic subalgebra $\mathbb{H} \subset \mathbb{O}$ determined by k and ℓ, we thus find one massive spin-$\frac{1}{2}$ particle at rest, with two spin states, namely

$$e_\uparrow = \begin{pmatrix} 1 \\ k \end{pmatrix}, \qquad e_\downarrow = \begin{pmatrix} -k \\ 1 \end{pmatrix}. \tag{14.88}$$

Both of these spinors (of course) have the same momentum, namely

$$e_\uparrow e_\uparrow^\dagger = e_\downarrow e_\downarrow^\dagger = \begin{pmatrix} 1 & -k \\ k & 1 \end{pmatrix} \tag{14.89}$$

corresponding to a particle at rest (with $m = 1$). The corresponding antiparticles are obtained by replacing k by $-k$ (and changing the sign in (14.80)).

We also find one massless spin-$\frac{1}{2}$ particle involving k, namely

$$\nu_z = \begin{pmatrix} 0 \\ k \end{pmatrix} \tag{14.90}$$

whose momentum is

$$\nu_z \nu_z^\dagger = \begin{pmatrix} 0 & 0 \\ 0 & 1 \end{pmatrix} \tag{14.91}$$

and which is therefore moving at the speed of light in the z-direction. This momentum-space state corresponds, as usual, to both a particle and its antiparticle. It is important to note that

$$\nu_{-z} = \begin{pmatrix} k \\ 0 \end{pmatrix} \tag{14.92}$$

has momentum

$$\nu_{-z} \nu_{-z}^\dagger = \begin{pmatrix} 1 & 0 \\ 0 & 0 \end{pmatrix} \tag{14.93}$$

and thus corresponds to a massless particle with the same helicity moving in the opposite direction, not to a different particle with the opposite helicity.

Each of the above states may be multiplied (on the *right*) by an arbitrary complex number without affecting its properties.

So far we have one massive particle with two helicity states, and one massless particle with a single helicity state. These are precisely the observed mass and helicity properties of a generation of leptons! We therefore interpret e as an electron, and ν as a neutrino.

We have so far been working with a particular quaternionic subalgebra. How many such subalgebras are there? We want to include our preferred

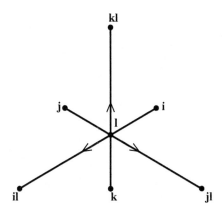

Fig. 14.1   Three quaternionic subalgebras of $\mathbb{O}$ which contain ℓ.

complex unit, ℓ, so that the reduction to four dimensions will work. So we are asking how many copies of $\mathbb{H}$ are there in $\mathbb{O}$ which contain a particular copy of $\mathbb{C}$. If we add the reasonable requirement that any two such subalgebras intersect *only* in $\mathbb{C}$, then there precisely three such subalgebras, labeled by i, j, and k, as shown in Figure 14.1. But there are precisely three generations of leptons!

The careful reader will have noticed that we have left out a massless particle, namely

$$\varnothing_z = \begin{pmatrix} 0 \\ 1 \end{pmatrix}. \tag{14.94}$$

This particle is *complex*, and has the opposite helicity from the neutrinos ν. As with the other massless states, this state describes both a particle and an antiparticle. Alone among the particles, this one does not contain k, and hence does not depend on the choice of identification of a particular quaternionic subalgebra $\mathbb{H}$ satisfying $\mathbb{C} \subset \mathbb{H} \subset \mathbb{O}$.

We emphasize this basic asymmetry in the octonionic description of particles. In a quaternionic description, we would have recovered the usual solutions of the Dirac description, namely a massive particle with two helicity states, and a massless particle with two helicity states. The fact that one helicity state is complex and the other quaternionic would seem to be simply a conventional choice. Over the octonions, however, there are three massive particles, each with two helicity states, three massless particles with just one helicity state, and a single, complex massless state with the opposite helicity; there are more massless particles of one helicity than the other.

In conclusion, there is room in this octonionic description for precisely the right number of leptons: exactly three generations, each containing massless particles having only a single helicity. These two properties, observed in nature, have yet to be explained theoretically.

But what about the extra particle, the $\varnothing_z$? In the picture presented here, it has no generation structure. The other neutrinos are hard enough to detect; it does not seem unreasonable to predict that this "sterile neutrino" would interact weakly if at all with ordinary matter. Is this dark matter?!

14.8   Cayley Spinors

We have argued in Sections 14.3–14.7 that the ordinary momentum-space (massless and massive) Dirac equation in $3+1$ dimensions can be obtained via dimensional reduction from the Weyl (massless Dirac) equation in $9+1$ dimensions. This latter equation can be written as the eigenvalue problem

$$\widetilde{\boldsymbol{P}}\psi = 0 \tag{14.95}$$

where $\boldsymbol{P}$ is a 2×2 octonionic Hermitian matrix corresponding to the 10-dimensional momentum and tilde again denotes trace reversal. The general solution of this equation is

$$\boldsymbol{P} = \pm\theta\theta^{\dagger}, \tag{14.96}$$

$$\psi = \theta\xi, \tag{14.97}$$

where θ is a 2-component octonionic vector whose components lie in the same complex subalgebra of $\mathbb{O}$ as do those of $\boldsymbol{P}$, and where $\xi \in \mathbb{O}$ is arbitrary. (Such a θ must exist since $\det(\boldsymbol{P}) = 0$.)

It is then natural to introduce a 3-component formalism; this approach was used by Schray [36, 37] for the superparticle. Defining

$$\Psi = \begin{pmatrix} \theta \\ \overline{\xi} \end{pmatrix} \tag{14.98}$$

we have first of all that

$$\mathcal{P} = \Psi\Psi^{\dagger} = \begin{pmatrix} \boldsymbol{P} & \psi \\ \psi^{\dagger} & |\xi|^2 \end{pmatrix} \tag{14.99}$$

so that Ψ combines the bosonic and fermionic degrees of freedom. Lorentz transformations can be constructed by iterating ("nesting") transformations of the form [5]

$$\boldsymbol{P} \mapsto \boldsymbol{M}\boldsymbol{P}\boldsymbol{M}^{\dagger}, \tag{14.100}$$

$$\psi \mapsto \boldsymbol{M}\psi, \tag{14.101}$$

which can be elegantly combined into the transformation

$$\mathcal{P} \mapsto \mathcal{M}\mathcal{P}\mathcal{M}^\dagger \tag{14.102}$$

with

$$\mathcal{M} = \begin{pmatrix} M & 0 \\ 0 & 1 \end{pmatrix}. \tag{14.103}$$

This construction in fact shows how to view $SO(9,1)$ as a subgroup of E_6; the rotation subgroup $SO(9)$ lies in F_4. It turns out that the Dirac equation (14.95) is equivalent to the equation

$$\mathcal{P} * \mathcal{P} = 0 \tag{14.104}$$

which shows both that solutions of the Dirac equation correspond to the Cayley plane and that the Dirac equation in ten dimensions admits E_6 as a symmetry group. Using the particle interpretation from Section 14.7 then leads to the interpretation of (part of) the Cayley plane as representing three generations of leptons.

We therefore refer to the 3-component octonionic column Ψ as a *Cayley spinor*. This name is a bit misleading, as Ψ is not a "spinor" in the classical sense of belonging to an appropriate representation of an orthogonal group (more precisely, of a spin group). However, Ψ does bear a similar relationship to the "vector" $\mathcal{P}$ as the true spinor ψ does to the vector $\boldsymbol{P}$.

We emphasize that not all 3-component octonionic columns are Cayley spinors. For Ψ to be a Cayley spinor, $\Psi\Psi^\dagger$ must satisfy the Dirac equation in the form (14.104), which, as we saw in Section 12.5, forces the components of $\Psi\Psi^\dagger$ to lie in a quaternionic subalgebra of $\mathbb{O}$. As in the above construction, we assume that the components of Ψ also lie in this quaternionic subalgebra.

We conclude with a bit of speculation. We refer to $\mathcal{P}$ as a "1-square", since it is a primitive idempotent (squares to itself and has trace 1). In general, we refer to decompositions of the form (13.161) as p-square decompositions, where p is the number of nonzero eigenvalues, and hence the number of nonzero primitive idempotents in the decomposition. If $\det(\mathcal{A}) \neq 0$, then $\mathcal{A}$ is a 3-square. If $\det(\mathcal{A}) = 0 \neq \sigma(\mathcal{A})$, then $\mathcal{A}$ is a 2-square. Finally, if $\det(\mathcal{A}) = 0 = \sigma(\mathcal{A})$, then $\mathcal{A}$ is a 1-square (unless also $\operatorname{tr}(\mathcal{A}) = 0$, in which case $\mathcal{A} = 0$). It is intriguing that, since E_6 preserves both the determinant and the condition $\sigma(\mathcal{A}) = 0$, E_6 therefore preserves the class of p-squares for each p. If, as argued above, 1-squares correspond to leptons, is it possible that 2-squares are mesons and 3-squares are baryons?

14.9 The Jordan Formulation of Quantum Mechanics

It is well-known that the Albert algebra, introduced in Section 11.2, is the only exceptional realization of the Jordan formulation of quantum mechanics [17, 29, 30, 38]; this is in fact how it was first discovered. We summarize this formalism here.

Recall that the *Cayley plane*, introduced in Section 12.5, consists of those Jordan matrices $\mathcal{V}$ which satisfy the restriction [13, 33]

$$\mathcal{V} \circ \mathcal{V} = \mathcal{V}, \qquad \operatorname{tr} \mathcal{V} = 1. \tag{14.105}$$

Elements of the Cayley plane correspond to projection operators in the Jordan formulation of quantum mechanics. As shown in Section 12.5, the conditions (14.105) force the components of $\mathcal{V}$ to lie in a *quaternionic* subalgebra of $\mathbb{O}$ (which depends on $\mathcal{V}$). Basic (associative) linear algebra then shows that each element of the Cayley plane is a primitive idempotent, and can be written as

$$\mathcal{V} = vv^{\dagger} \tag{14.106}$$

where v is a 3-component octonionic column vector, whose components lie in the quaternionic subalgebra determined by $\mathcal{V}$, and which is normalized by

$$v^{\dagger}v = \operatorname{tr} \mathcal{V} = 1. \tag{14.107}$$

For given $\mathcal{V}$, the vector v is unique up to a *quaternionic* phase. Furthermore, using (12.64) and its trace (13.76), it is straightforward to show that, for any Jordan matrix $\mathcal{B}$,

$$\mathcal{B} * \mathcal{B} = 0 \iff \mathcal{B} \circ \mathcal{B} = (\operatorname{tr} \mathcal{B}) \mathcal{B} \tag{14.108}$$

which agrees with (14.105) up to normalization, and which is therefore the condition that that $\pm\mathcal{B}$ can be written in the form (14.106) (without the restriction (14.107)). For any Jordan matrix satisfying (14.108), the normalization $\operatorname{tr} \mathcal{B}$ can only be zero if v, and hence $\mathcal{B}$ itself, is zero, so that

$$\mathcal{B} * \mathcal{B} = 0 = \operatorname{tr} \mathcal{B} \iff \mathcal{B} = 0 \tag{14.109}$$

since the converse is obvious.

We will need the useful identities (13.146) and (13.153) from Section 13.6 for any Jordan matrix $\mathcal{X}$, which can be verified by direct computation. We also have the remarkable fact that

$$\mathcal{A} * \mathcal{A} = 0 = \mathcal{B} * \mathcal{B} \implies (\mathcal{A} * \mathcal{B}) * (\mathcal{A} * \mathcal{B}) = 0 \tag{14.110}$$

which follows by polarizing (13.146),[11] and which ensures that the set of Jordan matrices satisfying (14.108), consisting of all real multiples of elements of the Cayley plane, is closed under the Freudenthal product.

In the Dirac formulation of quantum mechanics, a quantum mechanical state is represented by a *complex* vector v, often written as $|v\rangle$, which is usually normalized such that $v^\dagger v = 1$. In the Jordan formulation [17,29,30], the same state is instead represented by the Hermitian matrix $vv^\dagger$, also written as $|v\rangle\langle v|$, which squares to itself and has trace 1 (compare (14.105)). The matrix $vv^\dagger$ is thus the projection operator for the state v, which can also be viewed as a pure state in the density matrix formulation of quantum mechanics. The phase freedom in v is no longer present in $vv^\dagger$, which is uniquely determined by the state (and the normalization condition).

A fundamental object in the Dirac formalism is the probability amplitude $v^\dagger w$, or $\langle v|w\rangle$, which is not however measurable; it is the squared norm $|\langle v|w\rangle|^2 = \langle v|w\rangle\langle w|v\rangle$ of the probability amplitude which yields the measurable transition probabilities. One of the basic observations leading to the Jordan formalism is that these transition probabilities can be expressed entirely in terms of the Jordan product of projection operators, since

$$(v^\dagger w)(w^\dagger v) = \mathrm{tr}\,(vv^\dagger \circ ww^\dagger). \tag{14.111}$$

A similar but less obvious translation scheme also exists [17] for transition probabilities of the form $|\langle v|A|w\rangle|^2$, where A is a Hermitian matrix, corresponding (in both formalisms) to an observable, so that all *measurable* quantities in the Dirac formalism can be expressed in the Jordan formalism.

So far, we have assumed that the state vector v and the observable A are complex. But the Jordan formulation of quantum mechanics uses only the *Jordan identity*

$$(A \circ B) \circ A^2 = A \circ \left(B \circ A^2\right) \tag{14.112}$$

for two observables (Hermitian matrices) A and B. As shown in [30], the Jordan identity (14.112) is equivalent to power associativity, which ensures that arbitrary powers of Jordan matrices—and hence of quantum mechanical observables—are well-defined.

The Jordan identity (14.112) is the defining property of a *Jordan algebra* [29], and is clearly satisfied if the operator algebra is associative, which

---

[11]The necessary fact that $\det(\mathcal{A} + \mathcal{B}) = 0$ follows from the definition (11.15) of the determinant in terms of the triple product, the cyclic properties of the trace of the triple product, and the assumptions on $\mathcal{A}$ and $\mathcal{B}$.

will be the case if the elements of the Hermitian matrices A, B themselves lie in an associative algebra. Remarkably, the only further possibility is the Albert algebra of 3×3 octonionic Hermitian matrices [30, 38].[12]

14.10 The 3-Ψ Rule[13]

An essential ingredient in the construction of the Green–Schwarz superstring [39, 40] is the spinor identity

$$\epsilon^{klm}\gamma^\mu\Psi_k\overline{\Psi}_l\gamma_\mu\Psi_m = 0 \qquad (14.113)$$

for anticommuting spinors Ψ_k, Ψ_l, Ψ_m, where ϵ^{klm} indicates total antisymmetrization. This identity can be viewed as a special case of a Fierz rearrangement. An analogous identity holds for commuting spinors Ψ, namely

$$\gamma^\mu\Psi\overline{\Psi}\gamma_\mu\Psi = 0. \qquad (14.114)$$

It is well-known that the 3-Ψ Rule holds (only) for Majorana spinors in three dimensions, Majorana or Weyl spinors in four dimensions, Weyl spinors in six dimensions, and Majorana–Weyl spinors in ten dimensions. Thus, the Green–Schwarz superstring exists only in those cases [39, 40]. For the same reason, the only Yang–Mills theories with minimal supersymmetry occur in dimensions 3, 4, 6, and 10 (see for instance [41]).

Several authors [4, 42, 43] discussed the relationship of vectors and spinors in these dimensions to the four division algebras, $\mathbb{R}, \mathbb{C}, \mathbb{H}, \mathbb{O}$. As was shown by Fairlie and Manogue [44], the 3-Ψ Rule in all of these cases is equivalent to an identity on the γ-matrices, which holds automatically for the natural representation of the γ-matrices in terms of the four division algebras, corresponding precisely to the above four types of spinors. Manogue and Sudbery [45] then showed how to rewrite these spinor expressions in terms of 2×2 matrices over the appropriate division algebra, thus eliminating the γ-matrices completely.

We show here that the 3-Ψ Rule is in fact equivalent to an associator identity over the division algebras.

Let U, V, W be arbitrary octonionic vectors, that is, $1 \times n$ octonionic matrices. The *vector associator*

$$[U, V, W] = (UV^\dagger)W - U(V^\dagger W) \qquad (14.115)$$

---

[12]The 2×2 octonionic Hermitian matrices also form a Jordan algebra, but, even though the octonions are not associative, it is possible to find an associative algebra which leads to the same Jordan algebra [30, 31].

[13]The material in this section is adapted from the appendix of [22].

satisfies

$$[W, V, V] \equiv 0 \qquad (14.116)$$

which is established by direct computation using alternativity. Setting $W = V$ yields

$$[V, V, V] \equiv 0. \qquad (14.117)$$

Essentially the same argument establishes a similar formula when W is replaced by an octonionic scalar λ, namely

$$[\lambda, V, V] = 0 \qquad (14.118)$$

where we have implicitly defined yet another associator, namely

$$[\lambda, V, W] = (\lambda V^\dagger)W - \lambda(V^\dagger W). \qquad (14.119)$$

An interesting consequence of this result is the Hermitian conjugate relation

$$V^\dagger(V\overline{\lambda}) = (V^\dagger V)\overline{\lambda} \qquad (14.120)$$

which has applications to the right eigenvalue problems considered in Chapter 13.

We can polarize (14.116) to obtain

$$[U, V, W] + [U, W, V] \equiv 0 \qquad (14.121)$$

a special case of which is

$$[V, V, W] + [V, W, V] \equiv 0 \qquad (14.122)$$

obtained by setting $U = V$. A further special case of (14.121) is

$$[U, V, W] + [U, W, V] + [V, W, U]$$
$$+ [V, U, W] + [W, U, V] + [W, V, U] \equiv 0 \qquad (14.123)$$

obtained by adding cyclic permutations of (14.121), or, without requiring (14.116), by repeated polarization of (14.117).

In the commuting case, the (unpolarized) 3-Ψ Rule can be written in terms of a 2-component octonionic "vector" (really a spinor) V as [36]

$$\widetilde{(VV^\dagger)}\, V = 0 \qquad (14.124)$$

where

$$\widetilde{A} = A - \operatorname{tr} A \qquad (14.125)$$

corresponding to time reversal. It is straightforward to check that (14.124) holds by alternativity, or by using the identity

$$\operatorname{tr}(VV^\dagger) = \operatorname{tr}(V^\dagger V) = V^\dagger V \in \mathbb{R} \qquad (14.126)$$

to write

$$\widetilde{(VV^\dagger)}\,V = \left(VV^\dagger - V^\dagger V\right)V = (VV^\dagger)V - V(V^\dagger V) = [V,V,V] \quad (14.127)$$

and we see that the 3-Ψ Rule for commuting spinors is just (14.117).

Equivalently, we can rewrite the (polarized) 3-Ψ Rule in terms of 2-component "vectors" U, V, W as

$$\left(\widetilde{UV^\dagger} + \widetilde{VU^\dagger}\right)W + \left(\widetilde{VW^\dagger} + \widetilde{WV^\dagger}\right)U + \left(\widetilde{WU^\dagger} + \widetilde{UW^\dagger}\right)V = 0. \quad (14.128)$$

Using the identity

$$\operatorname{tr}\left(UV^\dagger + VU^\dagger\right) = \operatorname{tr}\left(V^\dagger U + U^\dagger V\right) = V^\dagger U + U^\dagger V \quad (14.129)$$

we have

$$\widetilde{UV^\dagger} + \widetilde{VU^\dagger} = \left(UV^\dagger + VU^\dagger\right) - \left(U^\dagger V + V^\dagger U\right) \quad (14.130)$$

where the last term is Hermitian and hence real. We thus have

$$\left(\widetilde{UV^\dagger} + \widetilde{VU^\dagger}\right)W = \left(UV^\dagger + VU^\dagger\right)W - W\left(U^\dagger V + V^\dagger U\right) \quad (14.131)$$

which implies that (14.128) is precisely the same as (14.123).

An analogous argument can be given for anticommuting spinors; this is essentially the approach used in [36]. Combining these results, the 3-Ψ Rule can be written without γ-matrices in terms of 2-component octonionic spinors ψ_α as

$$[\psi_1, \psi_2, \psi_3] \pm [\psi_1, \psi_3, \psi_2] + [\psi_2, \psi_3, \psi_1]$$
$$\pm [\psi_2, \psi_1, \psi_3] + [\psi_3, \psi_1, \psi_2] \pm [\psi_3, \psi_2, \psi_1] = 0 \quad (14.132)$$

for both commuting $(+)$ or anticommuting $(-)$ spinors. Both of these identities follow from the identity (14.117) applied to 2-component octonionic vectors, which is a special case of the more general identity (14.116) that holds for octonionic vectors of arbitrary rank.

Chapter 15

Magic Squares

15.1 The 2 × 2 Magic Square

In Section 9.2, we showed that the groups $SU(2, \mathbb{K})$ are

$$SU(2, \mathbb{R}) \cong SO(2), \tag{15.1}$$
$$SU(2, \mathbb{C}) \cong SO(3), \tag{15.2}$$
$$SU(2, \mathbb{H}) \cong SO(5), \tag{15.3}$$
$$SU(2, \mathbb{O}) \cong SO(9), \tag{15.4}$$

for $\mathbb{K} = \mathbb{R}, \mathbb{C}, \mathbb{H}, \mathbb{O}$, respectively (and where all congruences are local, that is, up to double cover). Similarly, in Section 9.3, we showed that the groups $SL(2, \mathbb{K})$ are

$$SL(2, \mathbb{R}) \cong SO(2, 1), \tag{15.5}$$
$$SL(2, \mathbb{C}) \cong SO(3, 1), \tag{15.6}$$
$$SL(2, \mathbb{H}) \cong SO(5, 1), \tag{15.7}$$
$$SL(2, \mathbb{O}) \cong SO(9, 1). \tag{15.8}$$

Finally, combining results from Sections 7.5, 8.2, and 9.4, we showed that

$$Sp(4, \mathbb{R}) \cong SU(2, \mathbb{H}') \cong SO(3, 2), \tag{15.9}$$
$$Sp(4, \mathbb{C}) \cong SU(2, 2) \cong SO(4, 2). \tag{15.10}$$

There is a pattern here, which correctly suggests that

$$Sp(4, \mathbb{H}) \cong SO(6, 2), \tag{15.11}$$
$$Sp(4, \mathbb{O}) \cong SO(10, 2), \tag{15.12}$$

as shown in Table 15.1. This table has some remarkable properties. Looking at the three columns labeled by $\mathbb{R}$, $\mathbb{C}$, and $\mathbb{H}$, there is a symmetry between

Table 15.1   Some groups over the division algebras.

$\mathbb{K}$	$\mathbb{R}$	$\mathbb{C}$	$\mathbb{H}$	$\mathbb{O}$
$SO(2,\mathbb{K})$	$SO(2)$	$SO(3)$	$SO(5)$	$SO(9)$
$SL(2,\mathbb{K})$	$SO(2,1)$	$SO(3,1)$	$SO(5,1)$	$SO(9,1)$
$Sp(2,\mathbb{K})$	$SO(3,2)$	$SO(4,2)$	$SO(6,2)$	$SO(10,2)$

rows and columns. Corresponding groups, such as $SO(3)$ and $SO(2,1)$, have the same dimension, and are in fact merely different real forms of the same group. Furthermore, comparing (15.3) with (15.9) leads us to suspect, again correctly, that the rows can perhaps be labeled with the split division algebras.

These symmetries also lead us to suspect that there should be a fourth row in Table 15.1, and a unified description of all of these groups at once. We provide such a description in Section 15.2.

15.2   The Geometry of $SU(2, \mathbb{K}' \otimes \mathbb{K})^1$

The groups in Table 15.1 are all orthogonal groups. What is the pattern to their signatures? Let's go ahead and label the rows with the split division algebras, as shown in Table 15.2. Let $\kappa = |\mathbb{K}| = 1, 2, 4, 8$, and for $\mathbb{K}'$ keep track separately of the number of positive-normed basis units, $\kappa'_+ = 1, 1, 2, 4$, and negative-normed basis units, $\kappa'_- = 0, 1, 2, 4$, with $\kappa'_+ + \kappa'_- = |\mathbb{K}'|$. Then the groups in the table are just $SO(\kappa + \kappa'_+, \kappa'_-)$! So we need a representation of $SO(\kappa + \kappa'_+, \kappa'_-)$.[2] The orthogonal groups are closely related to spinors and Clifford algebras, and we can translate this standard relationship into the language of division algebras.

We choose as a basis for the octonions $\mathbb{O}$ the unit elements

$$\{e_a, a = 1 \dots 8\} = \{1, i, j, k, k\ell, j\ell, i\ell, \ell\} \tag{15.13}$$

and similarly for the split octonions $\mathbb{O}'$ the unit elements

$$\{e_a, a = 9 \dots 16\} = \{1, I, J, K, KL, JL, IL, L\}. \tag{15.14}$$

In what follows, the *split quaternions* $\mathbb{H}'$ have basis $\{1, K, KL, L\}$, the *split complex numbers* $\mathbb{C}'$ have basis $\{1, L\}$, and of course $\mathbb{R}' = \mathbb{R}$.

---

[1]The material in this section is adapted from [12].
[2]Throughout this section, we continue to use $SO(\kappa + \kappa'_+, \kappa'_-)$ to refer to its double cover, the spin group $Spin(\kappa + \kappa'_+, \kappa'_-)$.

Table 15.2   The 2×2 magic square of Lie groups.

	$\mathbb{R}$	$\mathbb{C}$	$\mathbb{H}$	$\mathbb{O}$
$\mathbb{R}'$	SO(2)	SO(3)	SO(5)	SO(9)
$\mathbb{C}'$	SO(2, 1)	SO(3, 1)	SO(5, 1)	SO(9, 1)
$\mathbb{H}'$	SO(3, 2)	SO(4, 2)	SO(6, 2)	SO(10, 2)
$\mathbb{O}'$	SO(5, 4)	SO(6, 4)	SO(8, 4)	SO(12, 4)

We will work with the algebra $\mathbb{O}' \otimes \mathbb{O}$ and its subalgebras $\mathbb{K}' \otimes \mathbb{K}$, where $\mathbb{K}$ is any of the division algebras $\mathbb{R}$, $\mathbb{C}$, $\mathbb{H}$, $\mathbb{O}$, and $\mathbb{K}'$ any of their split versions. Elements of $\mathbb{O}' \otimes \mathbb{O}$ are simply (sums of) products of elements of $\mathbb{O}'$ and $\mathbb{O}$, such as "jK". We assume that $\mathbb{O}$ and $\mathbb{O}'$ commute with each other, that is, that

$$aA = Aa \qquad (15.15)$$

for all $a \in \mathbb{O}$, $A \in \mathbb{O}'$, and we will therefore usually write elements of $\mathbb{O}$ before elements of $\mathbb{O}'$.

15.2.1   *The Clifford Algebra* $C\ell(\kappa + \kappa'_+, \kappa'_-)$

We consider matrices of the form

$$\boldsymbol{X} = \begin{pmatrix} A & \overline{a} \\ a & -A^* \end{pmatrix} \qquad (15.16)$$

with $A \in \mathbb{K}'$ and $a \in \mathbb{K}$. Then $\boldsymbol{X}$ can be written as

$$\boldsymbol{X} = x^a \boldsymbol{\sigma}_a \qquad (15.17)$$

where there is an implicit sum over the index a, which takes on values between 1 and 16 as appropriate for the case being considered. When not stated otherwise, we assume $\mathbb{K}' = \mathbb{O}'$ and $\mathbb{K} = \mathbb{O}$, as all other cases are special cases of this one. Equation (15.17) defines the *generalized Pauli matrices* $\boldsymbol{\sigma}_a$, which are given this name because $\boldsymbol{\sigma}_1$, $\boldsymbol{\sigma}_2$, and $\boldsymbol{\sigma}_9$ are just the usual Pauli spin matrices.

We define the determinant of $\boldsymbol{X}$ in the obvious way, namely

$$\det \boldsymbol{X} = -AA^* - \overline{a}a = -|A|^2 - |a|^2. \qquad (15.18)$$

The collection of matrices of the form $\boldsymbol{X}$ forms a normed vector space of signature $(\kappa + \kappa'_+, \kappa'_-)$, using ($-1$ times) the determinant as the norm. We can therefore identify the resulting vector space with $\mathbb{R}^{\kappa+\kappa'_+,\kappa'_-}$.[3]

We now consider the 4×4 matrix

$$P = \begin{pmatrix} 0 & \boldsymbol{X} \\ \widetilde{\boldsymbol{X}} & 0 \end{pmatrix} = x^a \boldsymbol{\Gamma}_a, \tag{15.19}$$

where tilde represents trace reversal,

$$\widetilde{\boldsymbol{X}} = \boldsymbol{X} - \operatorname{tr}(\boldsymbol{X})\,\boldsymbol{I}, \tag{15.20}$$

and where the gamma matrices $\boldsymbol{\Gamma}_a$ are implicitly defined by (15.19). The only $\boldsymbol{\sigma}_a$ which are affected by trace reversal are those containing an imaginary element of $\mathbb{K}'$, which are imaginary multiples of the identity matrix. A straightforward computation using the commutativity of $\mathbb{K}$ with $\mathbb{K}'$ now shows that

$$\{\boldsymbol{\Gamma}_a, \boldsymbol{\Gamma}_b\} = 2g_{ab}\boldsymbol{I} \tag{15.21}$$

where $\boldsymbol{I}$ is the identity matrix and

$$g_{ab} = \begin{cases} 0 & a \neq b \\ 1 & 1 \leq a = b < 13\} \\ -1 & 13 \leq a = b \leq 16\} \end{cases} \tag{15.22}$$

is the metric. These are precisely the anticommutation relations necessary to generate a representation of the (real) *Clifford algebra* $C\ell(12,4)$ in the case of $\mathbb{O}' \otimes \mathbb{O}$, and $C\ell(\kappa + \kappa'_+, \kappa'_-)$ in general.

However, Clifford algebras are associative, so our algebra must also be associative. Since the octonions are not associative, neither are matrix groups over the octonions, at least not as matrix groups. The resolution to this puzzle is to always consider octonionic "matrix groups" as linear transformations acting on a particular vector space, and to use composition, rather than matrix multiplication, as the group operation. This construction always yields an associative group, since composition proceeds in a fixed order, from the inside out.

---

[3]There is a similar construction using matrices of the form

$$\boldsymbol{X}' = \begin{pmatrix} A & \bar{a} \\ a & A^* \end{pmatrix}$$

which have some remarkable properties. As a vector space, of course, the collection of such matrices provides another representation of $\mathbb{R}^{\kappa+\kappa'_+,\kappa'_-}$. But matrices of this form, unlike those of the form (15.16), close under multiplication; not only do such matrices satisfy their characteristic equation, but the resulting algebra is a Jordan algebra.

Let's start again. The Clifford identity that we need to verify is not (15.21) itself, but rather the action of (15.21) on (linear combinations of) products of gamma matrices, such as Γ_c. So we need to establish that

$$\Gamma_a\left(\Gamma_b\Gamma_c\right) + \Gamma_b\left(\Gamma_a\Gamma_c\right) = 2g_{ab}\Gamma_c. \qquad (15.23)$$

If the associator of the corresponding units vanishes, that is, if $[e_a, e_b, e_c] = 0$, then (15.23) reduces to (15.21), so we only need to consider cases where the associator is nonzero. But in this case an extra minus sign is required in each term on the left-hand side of (15.23) in order to factor out Γ_c, which changes nothing since both sides are 0. This argument establishes (15.23), and it is easy to see that the argument still holds if Γ_c is replaced by any product of gamma matrices. Thus, the linear transformations $\{\Gamma_a\}$ do indeed generate a Clifford algebra, namely $C\ell(\kappa + \kappa'_+, \kappa'_-)$.

15.2.2 The Orthogonal Group $SO(\kappa + \kappa'_+, \kappa'_-)$

Representations of Clifford algebras normally lead to representations of the corresponding orthogonal groups, using a well-known construction, so we expect $C\ell(\kappa + \kappa'_+, \kappa'_-)$ to lead us to a representation of $SO(\kappa + \kappa'_+, \kappa'_-)$; we show here that it does. However, our use of nonassociative division algebras in our representations requires some modifications to the standard construction.

In the associative case, the homogeneous quadratic elements of $C\ell(\kappa + \kappa'_+, \kappa'_-)$ act as generators of $SO(\kappa + \kappa'_+, \kappa'_-)$ via the map

$$P \longmapsto M_{ab}PM_{ab}^{-1} \qquad (15.24)$$

where

$$M_{ab} = \exp\left(-\Gamma_a\Gamma_b\,\frac{\theta}{2}\right) \qquad (15.25)$$

and $P = x^a\Gamma_a$ as above. So we consider first the case where $[e_a, e_b, e_c] = 0$, with a, b, c assumed to be distinct. Then the original Clifford identity (15.21) implies that

$$\Gamma_a\Gamma_a = \pm I, \qquad (15.26)$$

$$(\Gamma_a\Gamma_b)\Gamma_c = \Gamma_c(\Gamma_a\Gamma_b), \qquad (15.27)$$

$$(\Gamma_a\Gamma_b)\Gamma_b = (\Gamma_b)^2\Gamma_a = g_{bb}\Gamma_a, \qquad (15.28)$$

$$(\Gamma_a\Gamma_b)\Gamma_a = -(\Gamma_a)^2\Gamma_b = -g_{aa}\Gamma_b, \qquad (15.29)$$

$$(\Gamma_a\Gamma_b)^2 = -\Gamma_a^2\Gamma_b^2 = \pm I. \qquad (15.30)$$

With these observations, we compute

$$M_{ab} P M_{ab}^{-1} = \exp\left(-\Gamma_a \Gamma_b \frac{\theta}{2}\right) (x^c \Gamma_c) \exp\left(\Gamma_a \Gamma_b \frac{\theta}{2}\right). \tag{15.31}$$

From (15.26), if $a = b$, then M_{ab} is a real multiple of the identity matrix, which therefore leaves P unchanged under the action (15.24). On the other hand, if $a \neq b$, properties (15.27)–(15.29) imply that M_{ab} commutes with all but two of the matrices Γ_c. We therefore have

$$\Gamma_c M_{ab}^{-1} = \begin{cases} M_{ab}\Gamma_c, & c = a \text{ or } c = b \\ M_{ab}^{-1}\Gamma_c, & a \neq c \neq b \end{cases} \tag{15.32}$$

so that the action of M_{ab} on P affects only the ab subspace. To see what that action is, we first note that if $A^2 = \pm I$ then

$$\exp\left(A\alpha\right) = I\,c(\alpha) + A\,s(\alpha) = \begin{cases} I\cosh(\alpha) + A\sinh(\alpha), & A^2 = I \\ I\cos(\alpha) + A\sin(\alpha), & A^2 = -I \end{cases} \tag{15.33}$$

where the second equality serves to define the functions c and s. Inserting (15.32) and (15.33) into (15.31), we obtain

$$\begin{aligned} M_{ab}\left(x^a\Gamma_a + x^b\Gamma_b\right)M_{ab}^{-1} &= (M_{ab})^2\left(x^a\Gamma_a + x^b\Gamma_b\right) \\ &= \exp\left(-\Gamma_a\Gamma_b\theta\right)\left(x^a\Gamma_a + x^b\Gamma_b\right) \\ &= \left(I\,c(\theta) - \Gamma_a\Gamma_b\,s(\theta)\right)\left(x^a\Gamma_a + x^b\Gamma_b\right) \\ &= \left(x^a c(\theta) - x^b s(\theta)g_{bb}\right)\Gamma_a \\ &\quad + \left(x^b c(\theta) + x^a s(\theta)g_{aa}\right)\Gamma_b. \end{aligned} \tag{15.34}$$

Thus, the action (15.24) is either a rotation or a boost in the ab-plane, depending on whether

$$\left(\Gamma_a\Gamma_b\right)^2 = \pm I. \tag{15.35}$$

More precisely, if a is spacelike ($g_{aa} = 1$), then (15.24) corresponds to a rotation by θ from a to b if b is also spacelike, or to a boost in the a direction if b is timelike ($g_{bb} = -1$), whereas if a is timelike, the rotation (if b is also timelike) goes from b to a, and the boost (if b is spacelike) is in the negative a direction.

If $\mathbb{K}' \otimes \mathbb{K} = \mathbb{H}' \otimes \mathbb{H}$ (or any of its subalgebras), we're done: Since transformations of the form (15.24) preserve the determinant of P, it is clear from (15.18) that we have constructed $SO(6,2)$ (or one of its subgroups).

What about the nonassociative case? We can no longer use (15.27), which now contains an extra minus sign. A different strategy is needed.

If e_a, e_b commute, they also associate with every basis unit, that is

$$[e_a, e_b] = 0 \implies [e_a, e_b, e_c] = 0 \tag{15.36}$$

and the argument above leads to (15.34) as before. We therefore assume that e_a, e_b anticommute, the only other possibility; in this case, e_a, e_b are imaginary basis units that either both lie in $\mathbb{O}$, or in $\mathbb{O}'$. As before, we seek a transformation that acts only on the ab subspace. Consider the transformation

$$\boldsymbol{P} \longmapsto e_a \boldsymbol{P} e_a^{-1} \tag{15.37}$$

which preserves directions corresponding to units e_b that commute with e_a, and reverses the rest, which anticommute with e_a. We call this transformation a *flip* about e_a; any imaginary unit can be used, not just basis units. If we compose flips about any two units in the ab plane, then all directions orthogonal to this plane are either completely unchanged, or flipped twice, and hence invariant under the combined transformation. Such double flips therefore affect only the ab plane.

The rest is easy. We *nest* two flips, replacing (15.24) by

$$\boldsymbol{P} \longmapsto \boldsymbol{M}_2 \left(\boldsymbol{M}_1 \boldsymbol{P} \boldsymbol{M}_1^{-1} \right) \boldsymbol{M}_2^{-1} \tag{15.38}$$

where

$$\boldsymbol{M}_1 = -e_a \, \boldsymbol{I}, \tag{15.39}$$

$$\boldsymbol{M}_2 = \left(e_a \, c(\tfrac{\theta}{2}) + e_b \, s(\tfrac{\theta}{2}) \right) \boldsymbol{I}$$

$$= \begin{cases} \left(e_a \cosh(\tfrac{\theta}{2}) + e_b \, \sinh(\tfrac{\theta}{2}) \right) \boldsymbol{I}, & (e_a e_b)^2 = 1 \\ \left(e_a \cos(\tfrac{\theta}{2}) + e_b \, \sin(\tfrac{\theta}{2}) \right) \boldsymbol{I}, & (e_a e_b)^2 = -1 \end{cases} \tag{15.40}$$

give the two flips. Using the relationships

$$\left(e_a c(\alpha) + e_b s(\alpha) \right)^2 = e_a^2 c^2(\alpha) + e_b^2 s^2(\alpha) = e_a^2 = -g_{aa}, \tag{15.41}$$

$$e_a^2 c^2(\alpha) - e_b^2 s^2(\alpha) = -g_{aa} c(2\alpha), \tag{15.42}$$

$$2s(\alpha)c(\alpha) = s(2\alpha), \tag{15.43}$$

we now compute

$$\boldsymbol{M}_2 \left(\boldsymbol{M}_1 \left(x^a \boldsymbol{\Gamma}_a + x^b \boldsymbol{\Gamma}_b \right) \boldsymbol{M}_1^{-1} \right) \boldsymbol{M}_2^{-1} = \boldsymbol{M}_2 \left(x^a \boldsymbol{\Gamma}_a - x^b \boldsymbol{\Gamma}_b \right) \boldsymbol{M}_2^{-1}$$

$$= \left(e_a \, c(\tfrac{\theta}{2}) + e_b \, s(\tfrac{\theta}{2}) \right) \left(x^a \boldsymbol{\Gamma}_a - x^b \boldsymbol{\Gamma}_b \right) \left(e_a \, c(\tfrac{\theta}{2}) + e_b \, s(\tfrac{\theta}{2}) \right) \left(-g_{aa} \right)$$

$$= \left(x^a c(\theta) - x^b s(\theta) \, g_{aa} \, g_{bb} \right) \boldsymbol{\Gamma}_a + \left(x^b c(\theta) + x^a s(\theta) \right) \boldsymbol{\Gamma}_b. \tag{15.44}$$

and we have constructed the desired rotation in the ab plane.

We also have

$$\boldsymbol{\Gamma}_a \boldsymbol{\Gamma}_b = -e_a e_b \, \boldsymbol{I} \qquad\qquad ([e_a, e_b] \neq 0) \qquad (15.45)$$

so in the associative case (with e_a, e_b anticommuting), we have

$$\boldsymbol{M}_2 \boldsymbol{M}_1 = \left(g_{aa} c(\tfrac{\theta}{2}) + e_a e_b s(\tfrac{\theta}{2})\right) \boldsymbol{I} = g_{aa} \exp\left(-g_{aa} \boldsymbol{\Gamma}_a \boldsymbol{\Gamma}_b \, \frac{\theta}{2}\right) \qquad (15.46)$$

which differs from M_{ab} only in replacing θ by $-\theta$ (and an irrelevant overall sign) if $g_{aa} = -1$. In other words, the nested action (15.38) does indeed reduce to the standard action (15.24) in the associative case, up to the orientations of the transformations. In this sense, (15.38) is the nonassociative generalization of the process of exponentiating homogeneous elements of the Clifford algebra in order to obtain rotations in the orthogonal group.

We therefore use (15.24) if e_a and e_b commute, and (15.38) if they don't. Since both of these transformations preserve the determinant of $\boldsymbol{P}$, it is clear from (15.18) that we have constructed $\mathrm{SO}(\kappa + \kappa'_+, \kappa'_-)$ from $C\ell(\kappa + \kappa'_+, \kappa'_-)$.

15.2.3   *The Group* $\mathrm{SU}(2, \mathbb{K}' \otimes \mathbb{K})$

So far we have considered transformations of the form (15.24) and (15.38) acting on $\boldsymbol{P}$. In light of the off-diagonal structure of the matrices $\{\boldsymbol{\Gamma}_a\}$, we can also consider the effect these transformations have on $\boldsymbol{X}$. First, we observe that trace-reversal of $\boldsymbol{X}$ corresponds to conjugation in $\mathbb{K}'$, that is,

$$\widetilde{\sigma_a} = \sigma_a^*. \qquad (15.47)$$

The matrices $\boldsymbol{\Gamma}_a \boldsymbol{\Gamma}_b$ then take the form

$$\boldsymbol{\Gamma}_a \boldsymbol{\Gamma}_b = \begin{pmatrix} \sigma_a \sigma_b^* & 0 \\ 0 & \sigma_a^* \sigma_b \end{pmatrix} \qquad (15.48)$$

and, in particular,

$$\exp\left(\boldsymbol{\Gamma}_a \boldsymbol{\Gamma}_b \, \frac{\theta}{2}\right) = \begin{pmatrix} \exp\left(\sigma_a \sigma_b^* \, \frac{\theta}{2}\right) & 0 \\ 0 & \exp\left(\sigma_a^* \sigma_b \, \frac{\theta}{2}\right) \end{pmatrix}, \qquad (15.49)$$

so we can write

$$\exp\left(-\boldsymbol{\Gamma}_a \boldsymbol{\Gamma}_b \, \frac{\theta}{2}\right) \boldsymbol{P} \exp\left(\boldsymbol{\Gamma}_a \boldsymbol{\Gamma}_b \, \frac{\theta}{2}\right)$$
$$= \begin{pmatrix} 0 & \exp\left(-\sigma_a \sigma_b^* \, \frac{\theta}{2}\right) \boldsymbol{X} \exp\left(\sigma_a^* \sigma_b \, \frac{\theta}{2}\right) \\ \exp\left(-\sigma_a^* \sigma_b \, \frac{\theta}{2}\right) \widetilde{\boldsymbol{X}} \exp\left(\sigma_a \sigma_b^* \, \frac{\theta}{2}\right) & 0 \end{pmatrix}.$$
$$(15.50)$$

The 4×4 action (15.24) acting on $\boldsymbol{P}$ is thus equivalent to the action

$$\boldsymbol{X} \longmapsto \exp\left(-\sigma_a \sigma_b^* \frac{\theta}{2}\right) \boldsymbol{X} \exp\left(\sigma_a^* \sigma_b \frac{\theta}{2}\right). \tag{15.51}$$

on $\boldsymbol{X}$.

Transformations of the form (15.38) are even easier, since each of $\boldsymbol{M}_1$ and $\boldsymbol{M}_2$ are multiples of the identity matrix $\boldsymbol{I}$. These transformations therefore act on $\boldsymbol{X}$ via

$$\boldsymbol{X} \longmapsto \boldsymbol{M}_2 \left(\boldsymbol{M}_1 \boldsymbol{X} \boldsymbol{M}_1^{-1}\right) \boldsymbol{M}_2^{-1} \tag{15.52}$$

where $\boldsymbol{M}_1$, $\boldsymbol{M}_2$ are given by (15.40), but with $\boldsymbol{I}$ now denoting the 2×2 identity matrix.

Since $\boldsymbol{X}$ is Hermitian with respect to $\mathbb{K}$, and since that condition is preserved by (15.51), we have realized $\mathrm{SO}(\kappa + \kappa'_+, \kappa'_-)$ in terms of (possibly nested) determinant-preserving transformations involving 2×2 matrices over $\mathbb{K}' \otimes \mathbb{K}$. This 2×2 representation of $\mathrm{SO}(\kappa + \kappa'_+, \kappa'_-)$ therefore deserves the name $\mathrm{SU}(2, \mathbb{K}' \otimes \mathbb{K})$.

15.2.4 *Magic Squares*

This construction is reminiscent of the *Freudenthal–Tits magic square* of Lie algebras, which is discussed further in Section 15.3. Indeed, Barton and Sudbery [9] discuss the restriction of the Freudenthal–Tits magic square to precisely the case considered here, albeit at the Lie algebra level. Their restricted magic square contains the 16 Lie algebras

$$\mathfrak{v}_2(\mathbb{K}_1, \mathbb{K}_2) = sa(2, \mathbb{K}_1 \otimes \mathbb{K}_2) \oplus \mathfrak{so}(\mathrm{Im}\ \mathbb{K}_1) \oplus \mathfrak{so}(\mathrm{Im}\ \mathbb{K}_2) \tag{15.53}$$

where $\mathbb{K}_1$, $\mathbb{K}_2$ are division algebras (or possibly their split cousins), $sa(n, \mathbb{K}_1)$ denotes the anti-Hermitian tracefree $n \times n$ matrices with elements in $\mathbb{K}_1$ and $\mathfrak{so}(\mathrm{Im}\ \mathbb{K}_1)$ denotes the isometry Lie algebra of the imaginary elements of $\mathbb{K}_1$. The algebras $\mathbb{R}$ and $\mathbb{C}$ have no such isometries, and

$$\mathfrak{so}(\mathrm{Im}\ \mathbb{H}) = \mathfrak{so}(3), \tag{15.54}$$

$$\mathfrak{so}(\mathrm{Im}\ \mathbb{O}) = \mathfrak{so}(7). \tag{15.55}$$

We have argued here that the group version of (15.53) is

$$V_2(\mathbb{K}_1, \mathbb{K}_2) = \mathrm{SU}(2, \mathbb{K}_1 \otimes \mathbb{K}_2) \tag{15.56}$$

that is, that the groups in this 2×2 magic square deserve to be called $\mathrm{SU}(2, \mathbb{K}_1 \otimes \mathbb{K}_2)$. We have presented the construction above only for the "half-split" case, with $\mathbb{K}_1 = \mathbb{K}'$ and $\mathbb{K}_2 = \mathbb{K}$, which, as already noted, has interesting applications to physics, but the other two cases ("non-split" and "double-split") are similar.

Table 15.3   The 3×3 magic square of Lie groups.

	$\mathbb{R}$	$\mathbb{C}$	$\mathbb{H}$	$\mathbb{O}$
$\mathbb{R}'$	SO(3)	SU(3)	Sp(3)	F_4
$\mathbb{C}'$	SL(3,$\mathbb{R}$)	SL(3,$\mathbb{C}$)	$A_{5(-7)}$	$E_{6(-26)}$
$\mathbb{H}'$	Sp(6,$\mathbb{R}$)	SU(3,3)	$D_{6(-6)}$	$E_{7(-25)}$
$\mathbb{O}'$	$F_{4(4)}$	$E_{6(2)}$	$E_{7(-5)}$	$E_{8(-24)}$

15.3   The 3×3 Magic Square

The *Freudenthal–Tits magic square* was originally given in terms of Lie algebras [33,46]; the version shown in Table 15.3 lists instead particular real forms of the corresponding Lie groups. Vinberg [47] later gave a symmetric parametrization of the Lie algebras $\mathfrak{v}_3(\mathbb{K}_1, \mathbb{K}_2)$ in this magic square in the form

$$\mathfrak{v}_3(\mathbb{K}_1, \mathbb{K}_2) = sa(3, \mathbb{K}_1 \otimes \mathbb{K}_2) \oplus \operatorname{der}(\mathbb{K}_1) \oplus \operatorname{der}(\mathbb{K}_2) \qquad (15.57)$$

where $\mathbb{K}_1$, $\mathbb{K}_2$ are division algebras (or possibly their split cousins), $sa(n, \mathbb{K}_1)$ denotes the anti-Hermitian tracefree $n \times n$ matrices with elements in $\mathbb{K}_1$ and $\operatorname{der}(\mathbb{K}_1)$ denotes the derivations of $\mathbb{K}_1$, which are the Lie algebra version of automorphisms. The division algebras $\mathbb{R}$ and $\mathbb{C}$ have no derivations, and

$$\operatorname{der}(\mathbb{H}) = \mathfrak{so}(3), \qquad (15.58)$$

$$\operatorname{der}(\mathbb{O}) = \mathfrak{g}_2. \qquad (15.59)$$

Thus, the only change in the Vinberg construction between the 2×2 and 3×3 cases is the use of $\mathfrak{g}_2$ rather than $\mathfrak{so}(7)$ in the octonionic case, a difference that can be attributed to the lack of triality in the 2×2 case.

This magic square again possesses a symmetry between rows and columns; corresponding groups are again different real forms of the same group. But the most remarkable property of the Freudenthal–Tits magic square is that it contains four of the five exceptional Lie groups—and the fifth, G_2, appears implicitly as part of the Vinberg construction.

The Vinberg construction also suggests that the groups in the Freudenthal–Tits magic square can be described as SU(3, $\mathbb{K}' \otimes \mathbb{K}$), by analogy with the 2×2 case considered in Section 15.2. This construction is indeed possible, but is beyond the scope of this book.

Further Reading

- John C. Baez, The octonions, *Bull. Amer. Math. Soc.* **39**, pp. 145–205, (2002); http://math.ucr.edu/home/baez/octonions. *A great introduction. Written for mathematicians but not aimed at experts.*
- John H. Conway and Derek A. Smith, **On Quaternions and Octonions**, A K Peters, Ltd., Boston, 2003. *Short and sweet. Gorgeous mathematics, beautifully presented.*
- Feza Gürsey and Chia-Hsiung Tze, **On the Role of Division, Jordan, and Related Algebras in Particle Physics**, World Scientific, Singapore, 1996. *Aimed at physicists.*
- S. Okubo, **Introduction to Octonion and Other Non-associative Algebras in Physics**, Cambridge University Press, Cambridge, 1995. *Readable mathematics text.*
- Stephen L. Adler, **Quaternionic Quantum Mechanics and Quantum Fields**, Oxford University Press, New York, 1995. *Everything you ever wanted to know about generalizing quantum mechanics to the quaternions; advanced physics text.*
- Nathan Jacobson, **Structure and Representations of Jordan Algebras**, American Mathematical Society Colloquium Publications, **Vol. 39**, American Mathematical Society, Providence, 1968. *Everything you ever wanted to know about Jordan algebras; advanced mathematics text.*
- P. Lounesto, **Clifford Algebras and Spinors**, Cambridge University Press, Cambridge, 1997. *Introduction to Clifford algebras.*
- Boris Rosenfeld, **Geometry of Lie Groups**, Kluwer, Dordrecht, 1997. *Advanced mathematics text.*
- Richard D. Schafer, **An Introduction to Nonassociative Algebras**, Academic Press, New York, 1966 & Dover, Mineola NY, 1995. *Mathematics text suitable for advanced undergraduates.*

Bibliography

[1] Baylis, W. E. (1999). *Electrodynamics: A Modern Geometric Approach* (Birkhäuser, Boston).

[2] Crowe, M. J. (1967). *A History of Vector Analysis* (University of Notre Dame Press, Notre Dame), (reprinted by Dover Publications, 1985, 1994).

[3] Dray, T. (2012). *The Geometry of Special Relativity* (CRC Press/A K Peters, Boca Raton, FL).

[4] Sudbery, A. (1984). Division algebras, (pseudo)orthogonal groups and spinors, *J. Phys. A* **17**, pp. 939–955.

[5] Manogue, C. A. and Schray, J. (1993). Finite lorentz transformations, automorphisms, and division algebras, *J. Math. Phys.* **34**, pp. 3746–3767, eprint arXiv:hep-th/9302044.

[6] Wangberg, A. and Dray, T. (2009). Visualizing Lie subalgebras using root and weight diagrams, *Loci* **2**, available at http://mathdl.maa.org/mathDL/23/?pa=content&sa=viewDocument&nodeId=3287.

[7] Dray, T., Manogue, C. A. and Wilson, R. A. (2014). A symplectic representation of E_7, *Comment. Math. Univ. Carolin.* **55**, pp. 387–399, eprint arXiv:1311.0341.

[8] Freudenthal, H. (1954). Beziehungen der E_7 und E_8 zur Oktavenebene, I, *Proc. Kon. Ned. Akad. Wet. A* **57**, pp. 218–230.

[9] Barton, C. H. and Sudbery, A. (2003). Magic squares and matrix models of Lie algebras, *Adv. Math.* **180**, pp. 596–647.

[10] Kincaid, J. J. (2012). *Division Algebra Representations of* SO(4, 2), Master's thesis, Oregon State University, available at http://ir.library.oregonstate.edu/xmlui/handle/1957/30682.

[11] Kincaid, J. and Dray, T. (2014). Division algebra representations of SO(4, 2), *Mod. Phys. Lett. A* **29**, p. 1450128, eprint arXiv:1312.7391.

[12] Dray, T., Huerta, J. and Kincaid, J. (2014). The 2×2 Lie group magic square, *Lett. Math. Phys.* **104**, pp. 1445–1468.

[13] Harvey, F. R. (1990). *Spinors and Calibrations* (Academic Press, Boston).

[14] Penrose, R. and Rindler, W. (1984 & 1986). *Spinors and Space-Time* (Cambridge University Press).

[15] Manogue, C. A. and Dray, T. (1999). Octonionic Möbius transformations, *Mod. Phys. Lett. A* **14**, pp. 1243–1255, eprint arXiv:math-ph/9905024.

[16] Dündarer, R., Gürsey, F. and Tze, C.-H. (1986). Self-duality and octonionic analyticity of S^7-valued antisymmetric fields in eight dimensions, *Nucl. Phys. B* **266**, pp. 440–450.

[17] Gürsey, F. and Tze, C.-H. (1996). *On the Role of Division, Jordan, and Related Algebras in Particle Physics* (World Scientific, Singapore).

[18] Baez, J. C. (2002). The octonions, *Bull. Amer. Math. Soc.* **39**, pp. 145–205, available at http://math.ucr.edu/home/baez/octonions.

[19] Conway, J. H. and Smith, D. A. (2003). *On Quaternions and Octonions* (A K Peters, Natick, MA).

[20] Wilson, R. A. (2009). Octonions and the Leech lattice, *J. Algebra* **322**, pp. 2186–2190.

[21] Dray, T. and Manogue, C. A. (1998). The octonionic eigenvalue problem, *Adv. Appl. Clifford Algebras* **8**, pp. 341–364, eprint arXiv:math.RA/9807126.

[22] Dray, T., Janesky, J. and Manogue, C. A. (2000). Octonionic Hermitian matrices with non-real eigenvalues, *Adv. Appl. Clifford Algebras* **10**, pp. 193–216, eprint arXiv:math/0006069.

[23] Goldstine, H. H. and Horwitz, L. P. (1962). On a Hilbert space with nonassociative scalars, *Proc. Natl. Acad. Sci. USA* **48**, pp. 1134–1142.

[24] Ogievetskii, O. V. (1981). The characteristic equation for 3×3 matrices over octaves, *Russian Math. Surveys* **36**, pp. 189–190; *Uspekhi Mat. Nauk* **36**, pp. 197–198 (in Russian).

[25] Dray, T. and Manogue, C. A. (1998). Finding octonionic eigenvectors using *Mathematica*, *Comput. Phys. Comm.* **115**, pp. 536–547, eprint arXiv:math.RA/9807133.

[26] Okubo, S. (1999). Eigenvalue problem for symmetric 3×3 octonionic matrix, *Adv. Appl. Clifford Algebras* **9**, pp. 131–176.

[27] Dray, T., Janesky, J. and Manogue, C. A. (2000). Some properties of 3×3 octonionic Hermitian matrices with non-real eigenvalues, *Oregon State University preprint*, eprint arXiv:math/0010255.

[28] Dray, T. and Manogue, C. A. (1999). The exceptional Jordan eigenvalue problem, *Internat. J. Theoret. Phys.* **38**, pp. 2901–2916, eprint arXiv:math-ph/9910004.

[29] Jordan, P. (2933). Über die Multiplikation quantenmechanischer Größen, *Z. Phys.* **80**, pp. 285–291.

[30] Jordan, P., von Neumann, J. and Wigner, E. (1934). On an algebraic generalization of the quantum mechanical formalism, *Ann. Math.* **35**, pp. 29–64.

[31] Jacobson, N. (1968). *Structure and Representations of Jordan Algebras*, American Mathematical Society Colloquium Publications, Vol. 39 (American Mathematical Society, Providence).

[32] Schafer, R. D. (1966). *An Introduction to Nonassociative Algebras* (Academic Press, New York), (reprinted by Dover Publications, 1995).

[33] Freudenthal, H. (1964). Lie groups in the foundations of geometry, *Adv. Math.* **1**, pp. 145–190.

[34] Manogue, C. A. and Dray, T. (1999). Dimensional reduction, *Mod. Phys. Lett. A* **14**, pp. 99–103, eprint arXiv:hep-th/9807044.

[35] Dray, T. and Manogue, C. A. (2000). Quaternionic spin, in R. Abłamowicz and B. Fauser (eds.), *Clifford Algebras and Mathematical Physics* (Birkhäuser, Boston), pp. 21–37, eprint arXiv:hep-th/9910010.

[36] Schray, J. (1996). The general classical solution of the superparticle, *Class. Quantum Grav.* **13**, pp. 27–38.

[37] Schray, J. (1994). *Octonions and Supersymmetry*, Ph.D. thesis, Oregon State University, available at http://ir.library.oregonstate.edu/xmlui/handle/1957/35649.

[38] Albert, A. A. (1934). On a certain algebra of quantum mechanics, *Ann. Math.* **35**, pp. 65–73.

[39] Green, M. B. and Schwarz, J. H. (1984). Covariant description of superstrings, *Phys. Lett. B* **136**, pp. 367–370.

[40] Green, M. B., Schwarz, J. H. and Witten, E. (1987). *Superstring Theory* (Cambridge University Press, Cambridge).

[41] Baez, J. and Huerta, J. (2010). Division algebras and supersymmetry I, in R. S. Doran, G. Friedman and J. Rosenberg (eds.), *Superstrings, Geometry, Topology, and C* Algebras* (American Mathematical Society, Providence), pp. 65–80, eprint arXiv:0909.0551.

[42] Evans, J. M. (1988). Supersymmetric Yang–Mills theories and division algebras, *Nucl. Phys. B* **298**, pp. 92–108.

[43] Kugo, T. and Townsend, P. (1983). Supersymmetry and the division algebras, *Nucl. Phys. B* **221**, pp. 357–380.

[44] Fairlie, D. B. and Manogue, C. A. (1986). Lorentz invariance and the composite string, *Phys. Rev. D* **34**, pp. 1832–1834.

[45] Manogue, C. A. and Sudbery, A. (1989). General solutions of covariant superstring equations of motion, *Phys. Rev. D* **40**, pp. 4073–4077.

[46] Tits, J. (1966). Algèbres alternatives, algèbres de Jordan et algèbres de Lie exceptionnelles, *Indag. Math.* **28**, pp. 223–237.

[47] Vinberg, E. B. (1966). A construction of exceptional Lie groups, *Trudy Sem. Vekt. Tenz. Anal.* **13**, pp. 7–9 (in Russian), abstract of talk given on 2/11/64.

Index

Printed in the United States
By Bookmasters